AF572373

AIR, WATER AND SOIL POLLUTION SCIENCE AND TECHNOLOGY

AIR QUALITY

ENVIRONMENTAL INDICATORS, MONITORING AND HEALTH IMPLICATIONS

Air, Water and Soil Pollution Science and Technology

Additional books in this series can be found on Nova's website under the Series tab.

Additional e-books in this series can be found on Nova's website under the e-book tab.

Environmental Science, Engineering and Technology

Additional books in this series can be found on Nova's website under the Series tab.

Additional e-books in this series can be found on Nova's website under the e-book tab.

AIR, WATER AND SOIL POLLUTION SCIENCE AND TECHNOLOGY

AIR QUALITY

ENVIRONMENTAL INDICATORS, MONITORING AND HEALTH IMPLICATIONS

ARTHUR HERMANS
EDITOR

New York

For permission to use material from this book please contact us:
Telephone 631-231-7269; Fax 631-231-8175
Web Site: http://www.novapublishers.com

NOTICE TO THE READER

Additional color graphics may be available in the e-book version of this book.

Library of Congress Cataloging-in-Publication Data

Air quality : environmental indicators, monitoring and health implications / editor, Arthur Hermans.
pages cm
Includes bibliographical references and index.
ISBN: 978-1-62808-259-3 (hardcover)
1. Air quality--Environmental aspects. 2. Environmental indicators. 3. Air quality--Measurement. 4. Air quality monitoring stations. 5. Air quality--Health aspects. I. Hermans, Arthur.
TD883.A47843 2013
363.739'22--dc23
2013021113

Published by Nova Science Publishers, Inc. † New York

Contents

PREFACE

This book discusses the environmental indicators, monitoring and health implications of air quality. Topics in this compilation include the application of a Mars-based regression model to the air quality study in Northern Spain; microbial air sampling; the impact of air pollution from the mining-metallurgical complex on the content of total sulfur in plant material and soil; ecotoxicity assessment of traffic-related airborne pollution; and air quality in occupied school buildings spaces in the south of Portugal.

Chapter 1 - Microorganisms found in the air may represent a hazard for several human activities. Hospitals, pharmaceutical plants, food processing plants, waste treatment plants, electronic and space components factories, spacecraft, agriculture, cultural heritage, and offices are all examples of activities where microorganisms can cause damage to people and objects, with significant economic consequences. Measuring microorganisms in the air is important when managing those environments. In particular, microbial air monitoring is useful to confirm the presence of biological agents, evaluate any possible exposure, validate successful risk abatement procedures, support investigations on disease outbreaks, and comply with legal requirements and guidelines where available. Monitoring is also a useful tool for general scientific research, quality assurance and educational purposes. No common standard procedures for microbial air sampling have been adopted so far and variations in sampling methods make comparing results from different studies difficult. Therefore, it is important that a basis of knowledge is built to support appropriate choices and to make monitoring a valid support for the management of biological risk.

To sample viable biological particles in the air, passive and active sampling methods can be used. Active methods work by suctioning a known volume of air and projecting it against a collecting surface, which can be solid or liquid. The concentration of microorganisms in the air is then measured and expressed in colony forming units per cubic metre (CFU/m^3). Several types of devices are commercially available. The choice of an active device will depend on the area being monitored and all the factors affecting the efficiency of the procedure. Passive sampling methods rely on the use of settle plates being exposed to air for a defined period of time. Results are expressed in CFU/plate/time. Passive methods measure the rate at which viable particles settle on surfaces. To select the most appropriate sampling method for a monitoring procedure, it is important that objectives are carefully considered. Special attention should be given to microbiological sampling procedures that include a written, well defined protocol for sample collection and culturing, analysis and interpretation of results, and corrective actions based on the observed results.

The microbial monitoring of air should be viewed as a tool to assess the effectiveness of preventive interventions aimed at reducing microbial contamination and to identify critical situations in the continuous process of quality improvement.

Chapter 2 - The aim of this research work is to build a regression model of air quality by using the multivariate adaptive regression splines (MARS) technique in the Gijón urban area (Northern Spain) at local scale. With the rapid increases in processing speed and memory of low-cost computers, it is not surprising that various advanced computational learning tools such as neural networks have been increasingly used for analyzing or modeling highly nonlinear multivariate engineering problems. However, neural networks have been criticized for its long training process since the optimal configuration is not known a priori. This research work explores the use of a nonparametric regression algorithm known as multivariate adaptive regression splines (MARS) which has the ability to approximate the relationship between the inputs and outputs, and express the relationship mathematically. In this sense, hazardous air pollutants or toxic air contaminants refer to any substance that may cause or contribute to an increase in mortality or serious illness, or that may pose a present or potential hazard to human health. To accomplish the objective of this study, the experimental dataset of nitrogen oxides (NO_x), carbon monoxide (CO), sulphur dioxide (SO_2), ozone (O_3) and dust (PM_{10}) were collected over 3 years (2006–2008) and they are used to create a highly nonlinear model of the air quality in the Gijón urban nucleus (Spain) based on the MARS technique. One objective of this model is to obtain a preliminary estimate of the dependence between primary and secondary pollutants in the

Gijón urban area at local scale. A second aim is to determine the factors with the greatest bearing on air quality with a view to proposing health and lifestyle improvements. The United States National Ambient Air Quality Standards (NAAQS) establishes the limit values of the main pollutants in the atmosphere in order to ensure the health of healthy people. They are known as criteria pollutants. Firstly, this MARS regression model captures the main insight of statistical learning theory in order to obtain a good prediction of the dependence among the main pollutants in the Gijón urban area. Secondly, the main advantages of MARS are its capacity to produce simple, easy-to-interpret models, its ability to estimate the contributions of the input variables, and its computational efficiency. Finally, on the basis of these numerical calculations, using the multivariate adaptive regression splines (MARS) technique, conclusions of this work are exposed.

Chapter 3 - The area of Bor and the surroundings (Eastern Serbia) have been known for exploitation and processing of sulfide copper ores for more than a hundred years. Long-term ore exploitation in the examined area has made a great impact on local environment. For the purpose of evaluating the influence of emission from the mining-metallurgical complex, biomonitoring has been conducted in the urban-industrial and rural zone. Total sulfur concentrations were determined in samples of deciduous (birch and linden) and evergreen (spruce and pine) trees, as well as in moss and edible parts of fruits and vegetables (apple, pear, string beans and potato).

Significantly higher concentrations of the total sulfur in parts of woody plant species have been detected at the sampling sites Town Park (birch leaves 5745 $\mu g\ g^{-1}$ dw) and Hospital (birch leaves 5668 $\mu g\ g^{-1}$ dw) in the urban-industrial zone, which has been confirmed with the enrichment factors (EF). Comparing foliar parts, enrichment with sulfur decreases in the following order: birch leaves > linden leaves > spruce needles. Significantly higher concentrations of sulfur (and hence EF) in moss samples confirm the reliability of moss as a bioindicator of air pollution. In the sampled edible parts of fruits in the rural areas, substantially lower concentrations of the total sulfur have been detected (197.3-308.6 mg kg^{-1} dw) compared to vegetables (359.0-1119.4 mg kg^{-1} dw). Correlations among the sampling sites, based on the concentrations of the total sulfur in plant parts and soil of linden, birch, spruce and pine show the closest relationship among the Town Park, Hospital and Slatina ($R^2 \geq 0.8$), which indicates a similar level of pollution in the urban-industrial and rural zone.

Chapter 4 - Traffic-related airborne pollution has been well characterised from (human) toxicological aspects, also, the role of plants in active/passive biomonitoring has been extensively investigated. On the contrary, much less information is available on the ecotoxicity of these emissions. Ecotoxicological tests (or bioassays) are controlled, reproducible tests in which ecological responses are determined quantitatively. These ecological responses are termed test end-points. Mortality as an ultimate (and easy-to-quantify) end-point is widely used, but other, sub-lethal endpoints exist such as impairment of growth/development, reproductive and/or biochemical dysfunctions, morphological abnormalities, or biochemical responses. Most authors agree that bioassays can be used in biomonitoring. Bioassays, however, can not only provide quantitative information on the toxic effect, but a clear concentration-response relationship can be established and concentration-response or stressor-response patterns can be analysed.

For ecotoxicity testing, the *Vibrio fischeri* bioluminescence inhibition test has been used almost exclusively. This chapter gives an overview of the basic principles of the bioassay, gives examples on its application for ecotoxicity assessment of airborne pollution, and also includes state-of-the art developments such as lux-based biosensors or the kinetic version of the test. Apart from this bioassay, other sporadically applied methods are described.

Chapter 5 - The aim of this chapter is to evaluate the air quality in occupied school buildings spaces in the South of Portugal. The results of this review article are obtained in school buildings, with Mediterranean environment, from Kindergartens to University. In this work the numerical and the experimental analysis are presented and discussed.

In the numerical simulation the softwares, that simulate the building thermal behaviour and the computational fluid dynamics, are used. In the building thermal behaviour software the mean indoor air quality inside all buildings occupied spaces is evaluated, while in the computational fluid dynamics numerical model the air quality inside each occupied space, and mainly in the respiration area, is evaluated in detail. The building thermal behaviour numerical model considers the building opaque and transparent bodies, while the computational fluid dynamics numerical model considers, in each space, the occupants and the indoor bodies.

In the experimental measurements the carbon dioxide is evaluated. The carbon dioxide, using the tracer gas decreasing concentration, is used to evaluate the internal air quality, using the air exchange rate, the age of the air or the airflow rate.

The numerical and experimental techniques are used to evaluate the internal air quality level in spaces occupied by children and students, in the South of Portugal. The obtained results are used to develop techniques and methodologies in order to increase the internal air quality in these kinds of occupied spaces. In the study the natural and crossed ventilation, in cold and warm thermal conditions, are evaluated.

In: Air Quality
Editor: Arthur Hermans

ISBN: 978-1-62808-259-3

Chapter 1

MICROBIAL AIR SAMPLING

Cesira Pasquarella[1,*], Elisa Saccani[1], Manuela Ugolotti[2], Carlo Signorelli[1] and Roberto Albertini[3]

[1]Department of Biomedical, Biotechnological and Translational Sciences, University of Parma, Italy
[2]Hospital Hygiene Unit, University Hospital of Parma, Italy
[3]Department of Clinical and Experimental Medicine, University of Parma, Italy

ABSTRACT

Microorganisms found in the air may represent a hazard for several human activities. Hospitals, pharmaceutical plants, food processing plants, waste treatment plants, electronic and space components factories, spacecraft, agriculture, cultural heritage, and offices are all examples of activities where microorganisms can cause damage to people and objects, with significant economic consequences. Measuring microorganisms in the air is important when managing those environments. In particular, microbial air monitoring is useful to confirm the presence of biological agents, evaluate any possible exposure, validate successful risk abatement procedures, support investigations on disease outbreaks, and comply with legal requirements and guidelines where available. Monitoring is also a

* Corresponding author: Cesira Pasquarella, Department of Biomedical, Biotechnological and Translational Sciences, University of Parma, Via Volturno, 39, 43125 PARMA, Italy, email: ira.pasquarella@unipr.it.

useful tool for general scientific research, quality assurance and educational purposes. No common standard procedures for microbial air sampling have been adopted so far and variations in sampling methods make comparing results from different studies difficult. Therefore, it is important that a basis of knowledge is built to support appropriate choices and to make monitoring a valid support for the management of biological risk.

To sample viable biological particles in the air, passive and active sampling methods can be used. Active methods work by suctioning a known volume of air and projecting it against a collecting surface, which can be solid or liquid. The concentration of microorganisms in the air is then measured and expressed in colony forming units per cubic metre (CFU/m^3). Several types of devices are commercially available. The choice of an active device will depend on the area being monitored and all the factors affecting the efficiency of the procedure. Passive sampling methods rely on the use of settle plates being exposed to air for a defined period of time. Results are expressed in CFU/plate/time. Passive methods measure the rate at which viable particles settle on surfaces. To select the most appropriate sampling method for a monitoring procedure, it is important that objectives are carefully considered. Special attention should be given to microbiological sampling procedures that include a written, well defined protocol for sample collection and culturing, analysis and interpretation of results, and corrective actions based on the observed results.

The microbial monitoring of air should be viewed as a tool to assess the effectiveness of preventive interventions aimed at reducing microbial contamination and to identify critical situations in the continuous process of quality improvement.

1. Background

Microorganisms found in the air may represent a hazard for several human activities [1-59]. Hospitals, pharmaceutical plants, food processing plants, waste treatment plants, electronic and space component factories, spacecraft, agriculture, cultural heritage, and offices are all examples of activities where microorganisms can cause damage to people and objects, with significant economic consequences. For example, in the pharmaceutical and food industry, products can be damaged, with a potential risk for consumers; in hospitals, air microbial contamination is associated with the risk of infections, especially in immunocompromized patients; on cultural heritage sites, microorganisms can affect cultural property and may become a danger for operators and visitors. In order to reduce air microbial contamination to levels that are appropriate for contamination-sensitive activities, cleanrooms are built. A cleanroom is a room in which the concentration of airborne particles is controlled, and which is constructed and used in a manner to minimize the introduction, generation and retention of particles inside the room, and in which other relevant parameters, such as temperature, humidity and pressure are controlled as necessary. The ISO 14644-1 standard classifies cleanrooms according to the concentration of airborne particles, regardless of their biological or physical nature [60]. However, particularly where risks are associated with the presence of microorganisms, it is important to evaluate the microbiological quality of air. As Möller wrote, "everybody working in cleanrooms, or, generally speaking, when hygiene is important, must make reliable measurements of airborne microorganisms" [61].

Air microbiological sampling has two main objectives: to confirm the presence of microorganisms and to evaluate the possible exposure of objects and human subjects to microbiological hazards. In particular, sampling is useful when investigating a disease outbreak in which air might be implicated in the transmission, when assessing the microbial contamination associated with the use of tools, devices, procedures or to specific potentially hazardous situations (e.g. construction or renovation of buildings, bioterrorism), when validating successful risk abatement procedures, and when complying with legal requirements and guidelines. It is also a useful tool for general scientific research, quality assurance and educational purposes.

Air sampling started soon after microbiology was born. At first samples were collected not so much to assess the concentration of microbes in air but to prove the germ theory of diseases. Both Pasteur in 1861, and Tyndall in 1882, estimated the number of bacteria in the air by introducing a known volume of air into sterile containers and counting the number of containers that became infected [62,63]. In 1881, Koch introduced the settling method of air sampling by measuring the number of bacteria that settled onto a solid medium [64]. In 1933 Wells, in the US, invented the first active sampler [65] and in 1941 the active sampler Casella was produced in the UK [66]. The methods used to measure bioaerosol became important during the 2nd World War because of a threat of germ warfare. Military defense research therefore grew very fast in many countries, especially in the US, at Fort Detrick, and the UK, at Porton Down [61]. After the war, the production of antibiotics and other sterile medicines required the use of clean rooms in pharmaceutical industries. In hospitals, hip and knee replacement surgical areas were the first to require controlled-air environments to prevent surgical site infections [67]. The growing number of food factories and the longer distances that food had to travel increased the need for hygiene [1,3,20,25]. The need to reduce the microbial contamination of air promoted the research of sampling methods that could quantify airborne microorganisms. However, to date no common standard procedures for microbial air sampling have been set, so the choice of which sampling method should be used in a given situation and the interpretation of results are up to the single researcher. The consequent variability in sampling methods makes comparing results difficult. This issue is the subject of a large debate and many problems still remain unsolved [5,9,12,31,40,41,68-86].

2. Air Microbial Sampling

The objective of air microbiological sampling is to collect biological particles from the air in a manner which does not affect the ability to detect organisms (e.g. alteration in culturability or biological integrity). This ability depends on the physical and biological characteristics of the organisms and on the physical features of the sampling instruments. It is a very complex problem due to the qualitative heterogeneity of microorganisms and their dimensional diversity. The size of bacteria, for instance, varies generally from 0.2 to 20 μm and that of bacterial spores between 0.5 and 3 μm. Fungal spores vary between 1.5 and 15 μm and viruses between 0.02 and 0.30 μm. In the air it is possible to find single bacterial cells and spores, fungal spores, viruses or clumps of bacteria, spores or viruses. Nevertheless, in a confined indoor environment, where people represent the main source of contamination, microorganisms are generally carried by particles (mostly skin particles), which can have a few or hundreds of bacteria on them. These skin fragments can break up, and it is found that microbe-carrying particles in the air have an average equivalent diameter of 12 μm [61,87,88]. Microorganisms are also contained in the droplets ejected from the respiratory tract when talking, coughing, sneezing and breathing. The size and number of droplets ejected depends on how they are produced. Talking for 5 minutes and coughing each can produce 3,000 nuclei droplets; sneezing can generate approximately 40,000 droplets which then evaporate to particles in the size range of 0.5-12 μm [5]. Larger droplets, greater than 10 μm in size, settle in a short time, while droplets smaller than 10 μm can remain in the air long enough to be swept around by air currents and can travel considerable distances [87-89].

The construction of new buildings and the renovation and repair of existing ones, together with HVAC (Heating, Ventilation and Air Conditioning) systems, represent the main sources of fungal spores [5,56]. *Legionella* spp., which can be naturally found in water, can be diffused in the air when using taps, showers, bathtubs or through nebulization from an HVAC system [5].

An important reference regarding the evaluation of air microbial contamination in hospitals can be found in the study by Greene et al., who described the number of airborne contaminants encountered during an extensive survey of two hospitals [90] and characterized over 10,000 contaminants isolated during a 15-month period [91]. The mean count was about 20 CFU/ft^3, with a variability of 4.5 to 72.4 CFU/ft^3. Forty-eight percent of contaminants were associated with particles larger than 5 µm in diameter, 30 percent with particles between 2 and 6 µm in diameter, and 22 percent with particles smaller than 2 µm in diameter. About forty-two percent of isolated contaminants were Gram-positive cocci, 19.2 percent were Gram-positive rods, 14 percent were Gram-negative rods, 17.1 percent were molds, 2.2 percent were actinomycetes, 1.2 percent were yeasts and the remainder were assorted diphtheroids and cocco-bacillary types. The largest proportions of Gram-positive cocci and Gram-positive rods were found on particles larger than 6 µm in diameter, more than 84 percent of the Gram-negative rods were associated with particles smaller than 2 µm in diameter and more than 59 percent of molds were associated with particles 2 to 6 µm in diameter.

The relative humidity of the environment can have an important effect on the particles' aerodynamic size, vitality and survival, as well as on their aggregation [10,70]. The size and shape of aggregates are strongly influenced by relative humidity. The results of an experiment where air with a temperature of 22°C and a relative humidity of 8 or 98 percent was blown on fungal colonies to generate spores, showed that with a humidity of 98 percent, 45 percent of spores were in single form, while with a humidity of 8 percent single spores were less than 37 percent [70].

Bacterial and fungal spores are generally the prevalent microbial forms in the bioaerosol. These organisms can resist environmental stresses such as, for example, ultraviolet light, extreme temperatures and chemical pollutants. However, some microorganisms that resist drying such as *Staphylococcus* spp. and *Streptococcus* spp. can survive in a vegetative state for long periods and can travel considerable distances in air and still remain viable. They may also settle on surfaces and become airborne again as secondary aerosols during certain activities (e.g., floor sweeping and bed making).

The sampling of air requires the separation of particles from streamlines, which can be achieved using different types of forces. Microbiological samples can be collected from air using two methods: active sampling and passive sampling. Active sampling relies mainly on inertia, which is utilized using several different techniques: impact on solid substrate, impact on liquid substrate, filtration. Active sampling methods relying on electrostatic forces or thermal gradients are less common due to their difficulty. In passive sampling, both the force of gravity and inertia are used.

2.1. Active Sampling

Active volumetric sampling is based on the intake of a known volume of air that is projected against a collecting surface which can be solid or liquid. This technique provides a quantitative assessment of the microorganisms present in a given volume of air. The three main collection methods are impaction, impingement and filtration.

The impaction method on solid surface is the most commonly used for airborne microorganisms. This method separates particles from the airstream by utilizing their inertia to force their deposition onto a solid or semisolid collection surface. The collection surface is usually an agar medium for culture-based analysis or an adhesive-coated surface that can be analyzed microscopically. A great variety of impactor samplers are commercially available. They differ in their inlet characteristics and can be divided into slit samplers and sieve samplers. In the former, air is drawn through a single nozzle. An example of this is the Casella sampler, which collects microorganisms on a Petri dish containing culture medium that is placed on a rotating plate. These samplers can provide a temporal analysis of bioaerosol concentration. In sieve samplers, such as the Surface Air System (SAS), air is drawn through a plate with several nozzles and the particles impact on a surface located below the perforated plate. Sieve samplers can collect particles in several stages, as in cascade impactors such as the Andersen sampler. When particles are collected with the Andersen sampler, they pass through six nozzles with a decreasing diameter, so that at each stage progressively smaller particles are collected. Air enters through an opening located on the upper part of the sampler; it is then gradually accelerated by subsequent passages through a series of discs with progressively smaller nozzles, which give access to the various stages. The impact occurs on collecting surfaces placed below each perforated disc. The largest particles remain trapped in the first stages, while the smaller ones are progressively collected in the subsequent stages, thus creating a selection by size. An Andersen sampler positioned at the height of the subjects' respiratory system is capable of estimating their exposure to respirable particles, as each stage of the sampler simulates subsequent parts of the respiratory tract [8,92]. The use of sieve samplers can highlight the fact that fungal spores tend to travel as a single unit, while small bacteria are typically transported adherent to some supports (e.g. skin scales).

Centrifugal samplers also use inertial forces to separate the particles from the airstream, but in a radial geometry. The Reuter Centrifugal Sampler (RCS) is an example. Airborne particles are drawn in by an impeller and thereafter impact onto an agar-coated plastic strip lining the internal periphery of the impeller housing.

The main advantage of agar impactor samplers is the collection of microorganisms directly onto the culture medium, and agar plates can be incubated without further treatment. These samplers are very useful at low levels of airborne microorganisms and have, therefore, been used in hospitals and pharmaceutical plants [71]. However, their performance can be adversely affected by overloading when microbial concentrations are high, resulting in agar plates or strips containing overlapping colonies that are too numerous to count [92,93].

Liquid impingement uses mainly inertial forces for the removal of particles from the airstream, but also takes advantage of their distribution through the bubbles of the liquid. Incoming air is bubbled in a liquid physiological solution or liquid culture medium, which is then filtered, and the filter is transferred on agar. Impingers are commonly used for the retrieval of bioaerosol particles over a wide range of airborne particle concentrations. For analysis, the liquid sample can be concentrated by filtration or diluted by liquid addition. The sample can also be divided and several analysis methods can be used. However, impingers can be inefficient in collecting large numbers of hydrophobic particles such as fungal spores [94]. In addition, the collected liquid may evaporate, particularly in hot dry environments [95]. The collecting container must be sterilized after each sampling and must be handled very carefully since it can break easily. An example of an impinger sampler is the All-Glass Impinger-30 (AGI-30).

Filtration sampling achieves the separation of particles from the airstream by forcing air through a porous medium, usually nitrocellulose membrane filters with 0.8μm-diameter pores, or gelatine membranes with 3 μm pores. Inertial forces and other mechanisms such as interception, diffusion and electrostatic attraction, result in the collection of particles on the filter's surface. The simultaneous action of all of these forces also removes particles that are smaller than the filter's pores. Gelatine filters, which are currently the most commonly used, permit a wide variety of sample analysis. They blend perfectly with agar and dissolve when liquid media are used. Recently, gelatine membranes have been successfully used for searching endotoxins and viruses. Thanks to their solubility, they do not require the removal of the sampled material. However, when the ambient temperature exceeds 30°C and the relative humidity is higher than 90 percent, gelatine membranes may deteriorate. In these cases, nitrocellulose membranes should be used. An example of a sampler based on filtration is the Sartorius MD8 sampler.

Results obtained using active samplers are generally expressed as colony forming units per cubic metre (CFU/m^3). A CFU may arise from a single microorganism, from an aggregation of many microorganisms, or from one or more microorganisms that have been carried by a particle [58].

In case of impaction on a solid surface, a more accurate term would be *microbe-carrying particles* (MCP). Microorganisms are generally carried by particles (mostly skin particles), each of which carries many microorganisms, but when cultured generally gives rise to a single colony. This colony is then counted as a single microorganism, so there is a clear underestimation. Certain types of samplers, for instance liquid impingers, will completely or partially disrupt clumps. The resulting sample will therefore reflect the total number of individual organisms present in the air [5]. However, because of their low sampling volume and low rate of sampling, and because of their tendency to disrupt clumps of viable particles, impingement devices may not be suitable for sampling viable airborne particles [58].

2.1.1. Sampler Performance

The performance of bioaerosol samplers can be divided into physical and biological performance. The physical performance (physical efficiency) is the ability of the sampler to collect particles of different sizes, regardless of whether they are constituted by microorganisms or inert particles. The biological performance (biological efficiency) is the ability to collect biological particles without altering their viability. The physical parameters of different samplers inevitably affect their biological efficiency which also varies according to the characteristics of the microorganisms collected [24,72,84,95]. Physical parameters include inlet sampling efficiency and collection efficiency. Inlet sampling efficiency refers to the ability of the sampler inlet to extract particles from the environment without bias due to the size, shape or density of the particles. Collection efficiency is the ability of the sampler to remove particles from the airstream and transfer them to the collection medium. Depending on its density, size and shape, a microorganism may or may not be collected. When particles approach the collecting surface, the airflow breaks and flows around the obstacle. Large particles having enough inertia do not follow the variation of the airflow but keep moving in a straight line, thus impacting on the collecting surface. Smaller particles with a lower inertia tend to follow the stream of air and are not collected. As for the airflow rate, at low impact velocity, the particles tend to travel with the air streamlines and are not collected; at a higher critical impact velocity, microorganisms are collected by the agar by attaching themselves to the agar surface, but if does not penetrate sufficiently into the nutrient medium it can undergo drying; at a high impaction velocity the particle penetrates into the agar, but if the speed is too high the microbial cell may be damaged with a impaired biological efficiency [84,96]. A very important parameter for evaluating the physical efficiency of impactor samplers is the cutting diameter (cut-off size or d_{50}), which mainly depends on the characteristics of the impaction nozzle and the airflow rate. The d_{50} usually represents the particle diameter at which 50 percent of the particles are collected. However, due to the sharp cutoff characteristics of impactor samplers, their d_{50} is generally considered to be the diameter above which all particles are collected. In spite of being a very important characteristic, the cutting diameter has only been defined for a few samplers [24,92,97-99]. It should be provided for every sampler, so that a sampler with a d_{50} lower than the average size of the particles that will have to be sampled must be selected. For example, a sampler with a d_{50} of 4 μm should not be used to collect spores of *Aspergillus niger*, which have a diameter of 1-3 μm [24]. Particles carrying microorganisms present in indoor air generally have a size between 10 and 20 μm and can be collected efficiently with low speed impact sampling. However, air may also contain smaller particles that can cause problems when sampling at low speed. The cutting diameter can be reduced with the increase of the speed of suction, in order to sample particles with small size. As for the Andersen cascade sampler, the size separation of particles is carried out by varying the air velocity through the progressive decrease of the holes at each stage. The first stages have holes with a larger diameter, therefore larger particles with a sufficient inertia are collected. To provide greater inertia to smaller particles it is necessary to reduce the diameter of the holes; in this way the particles that are not retained by a stage for insufficient inertia have the probability of being retained in the subsequent stages. However, the impact velocity of the air hitting the collection medium must be a compromise between being high enough to allow the entrapment of viable particles down to approximately 1 μm, and being low enough to ensure viability of viable particles by avoiding mechanical damage or the break-up of clumps of microorganisms [58]. A loss of particles, with a consequent reduction in physical efficiency, can also be given to their deposition on the walls of the sampler.

Biological sampling efficiency is the ability of the sampler to collect the microorganisms preserving the biological integrity. The biological efficiency will be lower than physical efficiency, for a number of reasons, such as the survival of microorganisms during collection and the ability of the collection medium to support their growth [58]. Ideally, each sampler should collect all airborne microorganisms without altering the biological integrity.

The majority of analysis methods rely on culturing the collected microorganisms that are subjected to a variety of environmental stressors. In addition, the stress of impaction may injure the collected microorganisms, depending on their physiological characteristics. Therefore, sampling stresses, which can result in a reduction of the culturability of airborne microorganisms, are important in assessing the overall bioaerosol sampler performance; for example, filtration sampling has the disadvantage of viability loss of vegetative cells, in particular for *Escherichia coli*, presumably as result of desiccation during the sampling of nitrocellulose membranes [100]. However, it has been demonstrated that if using gelatine foam filters, this effect is not serious; in fact, gelatine membranes have the advantage of maintaining a moist environment during the sampling, therefore the collected microorganisms are not subjected to stress conditions that may damage them, and retain their vitality [101]. Scherwing et al. also demonstrated that a sampling of 8 consecutive hours on a membrane of gelatin does not alter the recovery of microorganisms [50]. Currently, the nitrocellulose membranes are typically used only for the sampling of bacterial and fungal spores, that resist drying, while for all other microbial forms gelatin membranes are recommended.

Stewart et al. studied the microbial stress due to the impaction of microorganisms onto an agar collection surface, of *Pseudomonas fluorescens* and *Micrococcus luteus*, also differentiating metabolic damage and structural damage using different culture media [96]. It was shown that an increase in the sampling flow rate and, therefore in impaction velocity, produce damages to the bacterial cell with a significant decline in the percentage of microorganisms recovered. The reduction of the collection efficiency can occur even when the velocity is low, because of the effect of insufficient embedding of the microorganism into the nutrient surface. The highest relative rate of recovery, approximately 51% for *Pseudomonas fluorescens* and 62%

for *Micrococcus luteus*, was observed at 40 m/sec (6.4 L/min) and 24 m/sec (3.8 L/min), respectively; higher speeds reduced the recovery of the cells for the impact damage, while a low speed reduced the recovery probably due to a low penetration of microorganisms into the collection medium. The *Micrococcus luteus* demonstrated less damage than the *Pseudomonas fluorescens*, suggesting the hardy nature of the Gram-positive strain versus the Gram-negative microorganism and that the stress from impact depends on the physiological characteristics of the microorganism. It has been observed that the recovery of microbial cells can be improved by the addition of certain compounds to the collection medium, such as osmoprotectants, which aid in the resuscitation of stressed or damaged cells.

Using impingement samplers, the high velocity does not seem to have an effect on the viability of microbial susceptible cells [95,102]. However, it is recommended that when using these samplers, a rapid sampling with high air flows should be used in order to prevent the reproduction of the microorganisms in the culture medium with a consequent overestimation. However, high air flows often produced on the surface of the liquid medium can form copious foams that can compromise the viability of the microorganisms and cause problems in the following filtration; to avoid this drawback, antifoam substances can be added to the liquid collection. Impingement samplers are particularly indicated to prevent drying of viable particles and are recommended for the collection of highly sensitive organisms such as algae and for the recovery of soluble substances (mycotoxins, antigens, endotoxin), as well as for the sampling of bacteria and viruses particularly sensitive to dehydration [71].

The sampling efficiency of the different samplers should be known and documented and, in particular, the sampler used should have a physical efficiency related to the aerodynamic diameter of the particles to be collected [57]. The ISO 14698-1, Appendix B, describes a technique for determining the collection efficiency of samplers used for counting airborne microorganisms. Manufacturers or third-party testing organizations will usually perform this evaluation. The manufacturer should provide instructions for the use of samplers, as well as information on their limitations [58].

2.1.2. Collection Time

Sample collection time is a parameter that influences the results. The sampling period must be sufficiently long to obtain a representative sample of the airborne microorganisms present without exceeding the upper quantitation limit of the sampler or causing losses in culturability of airborne organisms

[58,72]. In fact, counting or identification of the number of colonies is facilitated when there is an appropriate surface density of particles. Errors in counting colonies can occur when the number of these is too low or too high. If the number of colonies is too low there may not be representative and have a high variability; on the contrary, if the number of colonies is too high, it increases the possibility of error to an underestimation, because of the overlapping of the colonies and of the inhibitory effects of the microorganisms between them. For statistical accuracy, microbiologists traditionally enumerate only those plates with colony counts between 30 and 300 on a 100-mm-diameter plate [24,72,103]. The UNI EN 13098 indicates an appropriate density of the colonies for bacteria less than 5 colonies per cm^2 and for fungi less than 2 colonies per cm^2 [57].

The selection of sample collection time is complicated by the fact that bioaerosol concentrations may vary greatly over time. Where bioaerosol concentrations are supposed to vary greatly over time, several samples may be required to determine the average concentration, which may range from a relatively low bioaerosol concentration in *clean rooms* and *ultra-cleanrooms* to very high concentrations found in certain industrial and agricultural settings [72]. In each case the operator must assess the situation in that environment, choosing the type of instrument most suitable and often the path of compromise between the requirements of the representativeness of the sample and the acceptable analytical conditions.

Each sampler has different collection times, according to airflow and to the area of the collection surface. Nevalainen et al. calculated the optimal sampling time for five bioaerosol samplers by using the formula $t =(\delta)(A)/(C_a)(Q)$, where t is the sampling time, δ is the desired surface density on collection area, A is the area of sampling surface, C_a is the average expected bioaerosol concentration, and Q is the sampler flow rate [84].

It has been demonstrated that the increased sampling time results in decreased viability for aerosolized vegetative bacterial cells [72,104]. As air flows over the nutrient agar surface to a fan impactor, the agar may lose water content, resulting in a harder surface, and sampled microorganisms may rebound from the hardened surface and not be collected. If they are collected, they may be insufficiently embedded in the agar, with the consequent exposure to desiccation stresses. The rebound of particles on other particles already collected is also possible Thus, a doubling of sampling time may not result in a doubling of CFU; sampling several consecutive samples of short duration rather than a few samples for a long time interval is preferred. A sampling time no longer than three minutes for bacteria and no longer than 10 minutes for fungi is considered appropriate [105].

2.1.3. Sampling Volume

An important aspect of the volumetric active sampling is the volume of air to be sucked in order to have accurate results. In particular, in clean rooms it is necessary to aspirate a high volume of air. Pongiluppi pointed out that the volume of air sampled is often very limited compared to the overall quantity of air that enters in an ultraclean environment. For example, if in an environment of 100 m^3 with 35 air changes/hour, the volume of air sampled is 3.5 m^3, the percentage of air monitored corresponds to 0.12‰ out of a total of 28,000 m^3 supplied over eight-hours period [106]. One of the major limitations of active samplers is the very low amount of intake air [55]. In controlled environments where values not exceeding 3 CFU/m^3 are expected, several cubic meters of air should be taken, in order to have a sufficient level of precision and accuracy. Using a sampler with a flow rate of 80 L/min, the aspiration of one cubic meter may take 15 minutes; a longer period of time may be necessary in order to have a more representative sample [55]. There are samplers with a high flow rate, but in this case it is necessary to consider the possible interruption of the flow and the creation of turbulence which increases the possibility of contamination [55].

ISO 14698-1 norm states that in environments with high contamination, the impaction method and sample volume should be selected in way appropriate to achieving separate colonies, to allow the results to be interpreted. The sampler should have a sufficient flow rate to collect 1 m^3 in a reasonable time, without significant drying of the sampling medium, and an appropriate air impact speed to the culture medium [58]. In an ultraclean operating room, Whyte et al. recommended that at least 20 m^3 of air using the sampler Casella (700 L/min) be taken [107]. The Hospital Infection Society Working Party Report on microbiological commissioning and monitoring of operating theatre suites underlined that the larger the volume of air sampled, the greater the assurance of an accurate result; to detect 0.5 CFU/m^3, the volume sampled must be at least 2 m^3, preferably more, so high volume samplers are essential. It is not specified how much to sample microbiological commissioning and monitoring of operating theater suites; that will be determined by the microbial numbers being sought and the sampling equipment [108]. For conventionally ventilated theaters, sampling volume of around 1 m^3 is suggested, but greater than 0.25 m^3. Using volumes above this generates no substantial problems until either the colonies on the incubated plate get too crowded to enumerate accurately or the agar starts to dry out due to the volume of air passed over it. Using volumes lower than 1 m^3 may result in interpretational difficulties and tends to make the data more qualitative as the volume decreases.

Thio et al. demonstrated that air sampling performed during an outbreak of *Aspergillus* spp. failed to detect the presence of *Aspergillus* spp. when 160 L of air were sucked, while the sampling of 1,200 L of air allowed the isolation of *Aspergillus* spp. [109]. Therefore, they propose that 1,000 L (1 m^3) be the standard minimum when obtaining volumetric air samples to assess the healthcare environment for *Aspergillus*. In reviewing studies that specified the amount of air sampled during *Aspergillus* outbreak investigations, the authors found that samples ranged from 360 L to 1,000 L (1 m^3), with 80% of studies sampling less than 1,000 L. Already Streifel had suggested that for highly-filtered environments a minimum of 1,000 L of air should be sampled, since the likelihood of detecting 1 CFU/m^3 is reduced with volumes smaller than this [110]. And the CDC Guidelines for Environmental infection control in health-care facilities recommended the use of slit or sieve impactor samplers capable of collecting large volumes of air in short periods of time to detect low numbers of fungal spores in highly filtered areas [94,100,104,111].

2.1.4. Differences between the Samplers

The collection efficiency of any air sampler will vary depending on the device used. Moreover, the diversity of sampling time and the volume sampled justifies the variability of the results obtained using different samplers [15,72]. Several studies have shown that different samplers give different results when used simultaneously for the evaluation of the microbial air [94-143]. At present it is considered that none of the samplers available is appropriate to sample all types of bioaerosols and can be considered as the reference, even if the Andersen and the Impinger AGI-30, have been proposed for this purpose [72].

Some of the many studies that have compared the results of microbial contamination of air obtained with different samplers are mentioned below.

Jensen et al. compared the recovery of free bacteria aerosols of vegetative cells (*E.coli*) and bacterial endospores-formers (*B. subtilis*) by using eight samplers (Andersen -1, -2 and -6 stage, SAS, Mattson-Garvin, RCS, Gelman, and the AGI-30 as reference sampler). The results showed a wide variation in the collection efficiency of the samplers: SAS and RCS showed lower collection efficiency, both for *E. coli* and *B. subtilis*; the vegetative cells of *E. coli* were essentially killed by desiccation with the Gelman MF sampler; the endospore-forming, desiccation-resistant cells of *B. subtilis* emulate, in relative collection efficiencies, those of the AGI-30 sampler. This study indicates that the Andersen-6 STG and AGI-30 samplers were the samplers of choice for recovering aerosols of free bacteria and the authors recommend their use to ensure that an air sample representative of the entire particle size distribution of bioaerosols is collected [100].

Pitzurra et al. compared the results of microbial contamination obtained by the SAS sampler and those obtained by the RCS sampler into three different environments (operating room, corridor of an university institute and enclosures). In all environments, higher microbial values were obtained by using the RCS sampler compared with the SAS sampler [132]. Similar results were reported in other studies [120,125].

In a study performed by Ljungvist and Reinmuller, seven different impaction-type air samplers were evaluated: slit-to-agar samplers (BIAP, FH3, R2S); sieve samplers (Andersen -6 stage, MAS, SMA) and centrifugal sampler (RCS Plus). The lowest CFU values were obtained by the SMA and R2S sampler; in the first case the apparently low CFU values were probably due to the low impaction velocity (around 1m/s), which gives a high d_{50} value. On the other hand, while the higher impaction velocity (>50 m/s) will result in a lower d_{50} value, it might also produce greater impaction stress on the microorganisms [126].

Buttner and Stetzenbach evaluated the efficiency of four types of aerobiological samplers (Andersen -6 stage, SAS and Burkard) for the retrieval of fungal spores of *Penicillium chrysogenum*. The spore concentration was determined by counting the particles in the *Penicillium* spore size range (1.8-3.5 μm). The Andersen and Burkard samplers showed the highest collection efficiency and the Andersen sampler had the highest levels of sensitivity and repeatability [115].

Mehta et al. compared fungal counts using four samplers: Burkard, RCS Plus, SAS Super 90 e Andersen. The Andersen and Burkard samplers retrieved equivalent volumes of airborne fungi, while the SAS Super 90 and the RCS Plus measurements did not differ from each other, but were significantly lower than those obtained with the Andersen or Burkard samplers. In conclusion, the authors affirm that the Burkard device seems to be an excellent alternative to the Andersen unit for use in settings where the line current is not available and where the size and weight of the sampler must be minimized [128].

In a review of the literature on the sampling of spores of *Aspergillus* spp., which, however, did not consider the study of Mehta et al., the use of the SAS sampler was recommend as it combines the handling and the efficiency of collection [129]. Cooper et al. found a moderate agreement between the counts of *Aspergillus* spp. obtained by using the Reuter centrifugal air sampler and the Andersen N^2 air sampler, despite the reports of the inefficiency of the Reuter centrifugal air sampler in detecting spores less than 4 μm. However, the counts ranged between 0 and 5 UFC/m^3; therefore, the authors suggest the study should be repeated where the pathogenic fungal contamination is higher to draw conclusions on the validity of the Reuter centrifugal air sampler [144].

Bellin and Schillinger compared the efficiency of the Andersen-6 stage and the SAS in the evaluation of fungal contamination [113]. Data collected showed that the SAS sampler recovered lower levels of colony forming units than the Andersen 6 stage impactor. There was no statistically significant difference between the samplers when concentrations of *Cladosporium* were compared; however, the SAS sampler recovered about half the number of CFU for three other fungal categories i.e. *Aspergillus* and *Penicillium.* The authors concluded that *Cladosporium* is often the predominant fungal type, but where that is not the case, the airborne fungi concentrations obtained with the SAS sampler must be interpreted with caution.

Also, Lee et al. showed a lower collection efficiency of fungal spores by the SAS sampler compared to other samplers [123], while in a study conducted by Gangneux et al. no significant differences in fungal counts were observed between SAS and other devices [120].

Nesa et al. assessed the performance of four air samplers for fungal spore collection in the hospital environment (BioImpactor 100-08, Air Test Omega, Air Samplair Mas-100, Samplair) [131]. No significant difference in the efficiency of spore recovery was found between Air Test Omega, Mas-100 and BioImpactor, whereas Samplair was significantly less efficient. No significant difference was observed when comparing BioImpactor, which was selected to represent the three superior impactors, with a single-stage Andersen, MD8 and RCS High Flow. Therefore, the authors conclude that Air test Omega, Air Sampl'air Mas-100 and BioImpactor 100-08 are suitable for routine indoor evaluation of fungal contamination of air in hospitals. The results of this study were put into discussion by Ambroise et al. who compared the four samplers (Samplair, MAS-100, BioImpactor and RCS Plus). Besides highlighting the diversity of the fungal count obtained with the different samplers, they pointed out that the BioImpactor sampler, which was chosen by Nesa et al. as a reference standard, showed the highest variability in the results [111].

Verhhoeff et al. compared the results obtained with five samplers (Slit-to-agar sampler, Andersen -6 stage, SAS, RCS, Gelatine Filter sampler) in combination with four culture media for quantifying fungal contamination.

The coefficients of variation were high for all combinations; the highest count in terms of UFC/m^3 were obtained more frequently using the Slit-to-agar sampler and the Andersen in combination with DG18 (Dichloran 18% glycerol agar) and MEA (malt extract agar) media, even if the combination RCS/MEA showed the highest value. The highest number of species isolated was obtained by using the Slit-to-agar sampler and the Andersen sampler in combination with the DG18 medium [137]. The use of the DG18 was suggested also by Wu et al., who compared the fungal concentrations and genera based on enumeration from MEA and DG18 media in a hospital setting [138]. Results showed that DG18 medium was more effective than the MEA medium in collecting more fungal colonies in terms of both quantity and types of genera. The authors presume that such a finding might be attributed to the characteristics of DG18 in slowing colony growth so that the dominating genus will not over occupy the culture plate surface before the less competitive genus can fully develop.

Shintani et al. compared the counts of bacteria and fungi by using five samplers (SAS, Bio Samp MBS-1000, RCS High Flow, MAS-100, Millipore Air Tester) with two different culture media: SCDA (soybean casein digest agar) and SCDALP (soybean casein digest agar lecithin polysorbate) [134]. In this study a statistically significant difference in collection efficiency among air samplers was observed in the case of the SCDA medium, but not in the case of the SCDALP medium. This suggests that the difference between the air samplers is due to the inappropriate choice of the collecting media; therefore, on the basis of the results obtained, the authors recommend the use of the SCDALP medium with any sampler.

Tolchinski et al. evaluated the performance of two personal bioaerosol samplers based on a swirling cyclone with recirculating liquid film for monitoring the level of environmental and occupational airborne microorganisms. The results showed that these two newly developed personal bioaerosol samplers are capable of doing high efficiency, aerosol sampling (the cutoff diameters are around 0.7 μm for both samplers), and have proven to provide acceptable survival for the collected bioaerosols. By using an appropriate non-aqueous collection liquid, these two personal bioaerosol samplers should be able to permit continuous, long-period bioaerosol sampling with considerable viability for the captured bioaerosols [142].

Xu et al tested an Andersen six-stage sampler and a BioStage impactor with mineral-oil-spread agar plates in collecting indoor and outdoor bacterial and fungal aerosols [143]. Experimental results revealed that use of a mineral-oil-spread agar plate can substantially enhance culturable bioaerosol recoveries by Andersen type impactors (p-values,0.05). The recovery enhancement was shown to depend on bioaerosol size, type, sampling time and environment. In general, more enhancements (extra 20%) were observed for last stage of the Andersen six-stage samplers compared to the BioStage impactor for 10 min sampling. The work suggests that enhancements for fungal aerosols were primarily attributed to the reduced impaction stress, while for bacterial aerosols reduced impaction, desiccation and particle bounce played major roles. The developed technology can readily enhance the agar-based techniques including those high volume portable samplers for bioaerosol monitoring.

The variability of the results associated with the use of different samplers is certainly an aspect of great importance, especially when there are indications on the maximum acceptable levels of microbial contamination [2,15,16,19,23,33]. As emphasized by some authors, most of the rules refer to CFU/m^3, without specifying the type of sampler to be used [61,126].

The problem relating to the diversity of the results obtained with different samplers is underlined in the Guidelines on the prevention and safety in the operating rooms of the Lombardy Region [145]. For microbiological control of the operating room, the use of an active sampler is recommended; however, because of the variability of the results obtained with different samplers used at the same time, before starting the microbiological surveillance, a preliminary study is suggested in order to define the critical values. The CDC *Guidelines for environmental infection control in health-care facilities* include lists of problems not solved by the diversity of results obtained with the different samplers [5].

Errors in the results obtained among active samplers may also result from the lack of attention to the regular calibration of the device, which is a fundamental aspect that is often overlooked. As the d_{50} (cut size) of an active sampler depends on the flow rate through the nozzle/nuzzles, a decreased flow rate (e.g. due to a weakened battery) increases the d_{50}; thus, collection is shifted toward larger particle sizes. If some of the nozzles in a multinozzle impactor are plugged, the flow velocity through the remaining nozzles is increased, resulting in a lower d_{50} with increased collection and, possibly, increased stress on the microorganisms [72]. It is very important that the sampler used automatically signals any malfunctions.

The National Institute for Occupational Safety and Health (NIOSH) recommended all samplers should be calibrated before and after sampling to ensure the flow rate is within the manufacturer's specifications [24]. However, it was shown that the flow rate declared by the industry sometimes does not correspond to that measured; therefore, it is necessary to accurately verify that parameter. Macher and First found that the measured RCS air sampling rate was lower than the one published by the manufacturer (210±27 L/min *vs* 280 L/min) [99]. The authors affirm that incorrect calculations of the effective air sampling rate and disregarding the differences in collection efficiency among samplers can lead to false conclusions about the usefulness of samplers for measuring concentrations of airborne microorganisms.

Particular attention must be paid to the results obtained using the sieve samplers as the number of CFU must be reassessed according to the conversion table provided by the manufacturer in order to obtain the "most probable number" on the basis of the number of the nozzles in the head of the sampler. In fact, there is the statistical probability that a colony may be derived from more than a single colony forming unit passing through a hole; this probability increases in relation to the surface density on the culture plate; therefore, in case of a high air contamination level it becomes particularly relevant.

2.1.5. Sampler Choice

The wide variety of commercially available active samplers require a choice of the sampling technique and the type of sampler, which must be based on the careful consideration of the objective of the sampling and the information required, and the awareness of the limitations of the different samplers. Air samplers are designed to meet differing measurements; the sampler used must be suitable for the measurement to perform and must be compatible with the method of analysis [57]. There is no single sampler valid for the collection of all types of microorganisms, and no one type of sampler can be used to collect 100% of airborne microorganisms. It is, therefore, necessary to carefully select the appropriate sampler [5,24,58,72].

The CDC *Guidelines for environmental infection control in health-care facilities* lists the factors that must be considered when choosing an air sampler: viability and type of the microorganisms to be sampled; compatibility with the selected method of analysis; sensitivity of particle to sampling; assumed concentration and particle size; whether airborne clumps must be broken (i.e. total viable microorganisms count vs. particle count); the volume of air to be sampled and the length of time the sampler is to be continuously operated; background contamination; ambient conditions; sampler collection efficiency; effort and skills required to operate sampler; the availability and the cost of the sampler, plus back-up samplers in case of equipment malfunction; availability of auxiliary equipment and utilities (e.g., vacuum pump, electricity, and water)[5].

The ISO 14698-1 norm states that the sampling rate, the duration of sampling and the type of sampling device strongly influences the viability of the microorganisms that are collected; this is underlined because of the number and variety of microbial air sampling systems commercially available. The selection for a particular application should consider, as a minimum, the following factors: type and size of viable particles microorganisms to be sampled; sensitivity of the viable particles microorganisms to the sampling procedure; the expected concentration of microorganisms, ability to detect high or low levels of biocontamination; appropriate culture media; time and duration of sampling; ambient conditions in the environment being sampled; disturbance of unidirectional airflow by the sampling apparatus; sampler properties [58]. As for the air sampling in environments supplied by the unidirectional flow, it has been demonstrated that the aspiration will alter the air flow of the environment with inevitable errors on the counts of the collected particles. Therefore, for the accuracy of the results, air sampling in these environments requires an isokinetic sampling, in which the mean velocity of the air entering the sampling probe inlet is the same as the mean velocity of the unidirectional airflow at that location. For example, the sampler SAS was tested in the Low Speed Wind Tunnel by performing the following tests: air turbulence test, environmental test and exhaust air tests; these tests showed a minimal turbulence around the head of the sampler. The deviation of the unidirectional flow caused by the sampler is insignificant and the air expelled from the sampler does not contaminate the environment. It is not sucked in again by the sampler; it follows the flow of air. Even the sampler Sartorius MD8, which possesses a pump that sucks in the air to be processed, is declared suitable for isokinetic sampling of unidirectional air flow environments.

In reference to the characteristics of the sampler, the ISO 14698-1 recommends the consideration of the following factors: appropriate suction flow to low levels of contamination, appropriate impact speed; handling (weight and size) and ease of use, ease of cleaning, disinfection and sterilization; possibility of adding particles to biocontamination to be measured. The air in output by the sampler should not contaminate the environment that is controlled or be sucked in again by the sampler itself [58].

As for operating theatres, initially the recommendations were based on experiments carried out with the large volume (700 L/min) Casella slit sampler [33,107]. A lot of new types of samplers have appeared afterwards, and the

2002 Hospital Infection Society *Working Party* Report on Microbiological commissioning underlines that sampling air in operating theaters does not form a substantial section of this market, so many are not suitable. They list the most available microbial air samplers, together with the points needed to be considered in selecting an air sampler [108]: 1. Can it sample a suitable volume of air (more than 2 m^3 may be needed in ultraclean operating theaters) within a reasonable length of time (e.g. 10 minutes) or before dehydration effects may occur? 2. Can it be operated remotely (e.g. infra-red control or via extension lead)? 3. Is it easy to use and clean? 4. If unusual plates, strips or filters are used, how much do they cost? 5. Does it need to be physically close to the wound in a working ultraclean theater? 6. Has it been demonstrated to be reasonably effective in the published literature?

It can be said, however, that any sampler can be used, provided that it can sample volumes of air equal to or greater than 100 L, and has an impact velocity less than 20 m/sec [78]. The use of the same sampler and a standardization of the sampling parameters are needed to make comparisons between the results obtained in different studies [16,78].

2.1.6. Number of Samples

Multiple samples must be taken in order to more accurately determine the concentration and composition of the bioaerosol [72]. In fact, a single sample, at a specific point in time and space usually has limited value in the assessment of the airborne microorganisms in an environment of concern. Because of the temporal variability of the bioaerosol, the air environment not perfectly mixed, the sporadic aerosolization of microorganisms, the short period of time sampling and the small volume of air sampled, a negative result obtained from an air sample is not proof of a specific microorganism. The number of air samples required depends on the statistical methods used to analyze the data, on the length of the sample collection time, and on the variability of the bioaerosol. Pilot sampling studies should be performed in a particular environment to arrive at the final sampling design. Simultaneous paired samples are often taken because of the high degree of variability which has been demonstrated between data from paired samplers of the same type [72]. Verhoeff et al. showed high coefficients of variation in the values of fungal load in samples taken in succession or in parallel, and conclude that the presence of molds in an environment, in terms of the number of colonies and the number of species, cannot be determined based on a single sample [136]. Ambroise et al. underlined the diversity in microbial counts obtained from air samples taken in triplicate, even with different volumes of air [111].

Differences in microbial contamination values obtained from consecutive samplings were also highlighted by other authors [123,146].

The *Hospital Infection Society Working Party* Report on Microbiological Commissioning recommended that at least two samples be taken, per theater, as this lessens the possibility of technical errors that could interfere with successful control of a theater [108].

2.2. Passive Sampling

For the separation of particles from the airflow, passive sampling utilizes the force of gravity and the inertial mass. In a hypothetical environment in which air could be stopped, sampling would only depend on the force of gravity. However, since air in an environment is constantly in motion, the collection of particles also depends on their inertia. In particular, the more the air is moving, the more the influence of the inertial component increases compared to that of gravitational force.

Particles in the air settle faster as their size increases. In a static-air environment, with a temperature of 25°C, hypothetical spherical particles with a density of 1 and a diameter of 2 μm settle with a speed of 0.012 cm/s. If the diameter is increased to 10 μm, sedimentation speed will increase to 0.3 cm/s, while particles with a diameter of 30 μm settle at a speed of 2.7 cm/s [147]. A study carried out in pharmaceutical clean rooms showed a settling velocity for microorganism-carrying particles of 0.2 cm/s to 3 cm/s [86].

Passive sampling is performed using Petri dishes containing a nutrient medium that are left exposed to air for a given period of time. Results are expressed as CFU/plate/time. Settle plates with a diameter of 9 to 14 cm are generally used. However, in order to make results from plates of different sizes comparable, some authors have proposed that results be expressed as CFU/m^2/h [76]. The suggestion of ISO 14698-1 is to express the results as viable particles per 1 dm^2 per hour [58]. In active sampling, and also in passive sampling, we should refer to microorganism-carrying particles instead of colony forming units, since microorganisms are mostly carried by particles.

Passive sampling is the oldest method to assess the microbial contamination of air. Settle plates were first used by Louis Pasteur in his studies on the origin of life [62] and were adopted by Robert Koch in 1881 to assess the quality of indoor air [64]. Fisher proposed them for air sampling in hospitals and introduced the standard 1/1/1: at one meter from the ground, one meter from any significant physical obstacle, and for a one-hour exposure

[148]. This standard was subsequently improved with the definition of the Index of Microbial Air contamination (IMA), corresponding to the number of CFUs that are deposited on a 9 cm-diameter Petri dish containing nutrient agar, left open to the air for 1 hour, 1 meter above the floor, 1 meter from the wall, and the definition of threshold values to interpret the results [85]. In order to better comply with sampling parameters, an automatic passive air sampler was devised. The Sed-Unit device, developed by EMPA (Eidgenössische Materialsprufüngs and Forschungsantstalt) in St. Gallen, Switzerland, allows the correct positioning of the Petri dish and makes the measurement of the IMA easier and more accurate. A moving arm automatically opens and closes the plate. The machine can be operated with a programmable delay of 2 minutes to 24 hours. Once programmed, the Sed-Unit works independently of human operators, thus reducing the chances of microbial shedding from the operator's body.

Since the volume of air from which particles are collected is unknown, passive sampling is not a volumetric sampling. It does not measure the total number of viable particles in the air, but rather the rate at which particles settle on surfaces. Settle plates may therefore be used for a quantitative and qualitative assessment of the contamination of a surface due to airborne particles [58]. Settle plates provide a direct measurement of the amount and type of microorganisms that are likely to be deposited on a critical surface (e.g. surgical wound in operating theater, a medicament in pharmaceutical industries, a food product in food industries). This characteristic is considered an advantage when the researcher's focus is on microbial sedimentation [5,9,30,31,58,85,125,149].

As Whyte underlined, it is possible to have two environments with identical volumetric counts, but different rates of deposition of microorganism-carrying particles, hence with a different degree of contamination of the exposed surface and, consequently, a different level of risk [86]. Any critical surface exposed to the air is more or less subject to contamination by airborne microorganisms based on the following factors: surface area, air exposure time, and microbiological quality of air measured by volumetric sampling or settle plates. Knowing the CFU/m^3 value provided by volumetric samplers, the number of microorganisms settling on a critical surface can be estimated using the following formula: concentration of microorganisms in the air x deposition rate x surface area x exposure time. For example, Whyte et al. estimated at 169 the number of microorganisms that are directly deposited onto a wound with an area of 0.0124 m^2 during an operation lasting 110 minutes, assuming an airborne bacterial count of 413 CFU/m^3, an average particle size of 12 μm and a settling velocity of 0.3 m/min (413x0.3x0.0124x110) [150]. The difficulty of estimating the number of microorganisms that are deposited, using the equation shown, comes from the problem of knowing exactly the value of deposition velocity of the particles, which can be calculated by simultaneously using the volumetric sampling and the settle plates or by reference to published data [86]. Such difficulty may, however, be overcome through the use of settle plates. If a plate is left in the vicinity of the surface at risk, both the surface and the plate will be contaminated by the microorganisms in the same way, proportionally to the surface area and the time of exposure; the number of CFUs counted on the plate will, therefore, allow us to directly estimate the number of CFUs deposited on the surface at risk.

Whyte et al. have demonstrated that the number of microorganisms sedimented on a plate left exposed in the vicinity of the surgical wound was similar to the number of microorganisms obtained from the wound wash-out counts [150]. As a result of an experiment carried out in a pharmaceutical industry [107], over a range of airborne microbial concentrations (from <1 CFU/m^3 to 1100 CFU/m^3) and times during which the vial or ampoules were exposed (from 5 seconds to 20 minutes) it was shown that the estimated contamination rate calculated from settle plates counts was close to the actual contamination rate determined by broth fill [86]. The CDC consider the plates suitable for the sampling of bacteria in the vicinity of a medical procedure, during its development, or for a general measure of the microbial quality of the air [5].)

In a study by Humphrey et al. settle plates were used for the assessment of the contamination of the air of operating theaters in 49% of hospitals; however, the authors concluded by saying that even if the plates were cost-effective, they were not appropriate for this purpose [151]. Humphrey et al., in a later article, pointed out the still open debate concerning the method to be used for air microbial monitoring in operating theaters, and quoted a study by Friberg et al. [153] whose results suggested that the plates may have a role in assessing the microbiological quality of the operating theater because they reflect the bacterial load in the immediate vicinity of the surgical site and thus the risk of contamination of the wound itself. Such opinion is supported by other authors [31,73,75,81,85]. Where sampling is performed, to estimate the risk of microbial contamination of a surface is critical; settle plates placed in the immediate vicinity of the surface may represent an easy, cheap and useful method of control, also with the possibility of a continuous monitoring of the environment [86].

Several authors emphasize the lack of sensitivity and reproducibility of the settle plates so that they would not be suitable to provide representative quantitative and qualitative results [5,31,71,109,129,151]. As sedimentation is influenced in particular by the size and shape of the particles and by the movements of the surrounding air, large particles are more likely to be deposited, while the smaller particles may be excluded from the collection. However, the sensitivity of the method can be increased, lengthening the time of exposure and/or using settle plates with a larger diameter [86]. The European Guide to Good Manufacturing Practice for the production of sterile medicinal products (EC GMP) reports the recommended limits for microbiological monitoring of clean areas during operation considering an exposure of four hours [16]. It has been shown that the plates may be left exposed for several hours without a significant loss of microorganisms, caused by the impairment of the vitality due to dehydration of the agar.

Russell et al. showed [86,115,154,155] that the microbial count from an 8 hour exposure of a settle plate was not significantly different from the addition of the count obtained from 16 separate half-hour exposures. [154]. Whyte and Niven demonstrated that leaving a plate exposed for 24 hours in a still room, and for 6 hours in unidirectional flow, can have a 13% reduction in the water content having little affect on the viability of the microorganisms deposited (approximately 8%)[155]. Buttner and Stetzenbach showed that, unlike what occurs in the active sampling, in which a long sampling time is associated with a significant reduction in the recovery of vegetative cells, in the passive sampling an increase in the exposure time does not significantly affect that recovery [115]. However, attention must be paid to the thickness of the agar, because if the agar plates contain only a thin layer, appreciable dehydration may occur[155].

The duration of exposure of the plate is an important factor that qualifies the passive sampling [9,85,86,137]. For example, settle plates left open for an hour during a surgical operation of 5 hours give a measure of the sedimentation of the microorganisms for a time that corresponds to 20% of the entire operating session. This is long enough to record all possible variations that may affect the microbial air contamination, such as, for example, number of people present, variations in their activities, arrival of outside visitors, and door openings. Considering the fluctuation in microbial air contamination resulting from frequent aerosol production, some authors deem the use of settle plates more appropriate than active samplers in dental clinics, as they provide a measure of the contamination build-up during clinical activity [12,45].

Other advantages of passive sampling are the simplicity of performing [9,86,139], the economy [9,30,31,85], the easy availability of the plates sedimentation [9,31,85,86], the ability to use sterile plates, which can be placed very close to the surface at airborne risk [31,85], the possibility to perform samplings contemporaries [9,31,85], the ability to compare the results [31,85], the reproducibility [85,137], the possibility to use the plates in environments with unidirectional flow; without that the flow of air is discontinued and creates turbulence [8,47], the maintenance of biological organisms [9,85]. With regard to this last aspect, the CDC *Guidelines for environmental infection control in health-care facilities* underlines that because the survival of microorganisms during air sampling is inversely proportional to the velocity at which the air is taken into the sampler, one advantage of using a settle plate is its reliance on gravity to bring organisms and particles into contact with its surface, thus enhancing the potential for optimal survival of collected microorganisms [5].

Möller stresses that the use of the plates of sedimentation does not lead the problems related to the use of active samplers, such as those due to possible technical failures or difficulties in comparing the results [61].

3. Comparison between Active and Passive Sampling

Several studies have compared the values of the microbial counts obtained with the active and passive sampling. In some cases a significant correlation was found [40,76,137,139,156-165], in others the correlation was absent [76,161,163].

Analyzing the results obtained in two pharmaceutical cleanrooms during a variety of airborne microbial conditions, Whyte found a good correlation between volumetric counts obtained by using the Casella slit sampler and settle plates counts [86]. This correlation was the basis for the recommended limits expressed both as CFU/m^3 and as CFU/plate 9 cm Ø/4 hours of the EC GMP [16].

Friberg et al. evaluated the correlation between bacterial counts obtained by active and passive sampling in operating theaters equipped with a conventional air flow with the surgical team wearing cotton clothing or non-woven disposables. A correlation was observed in all monitored points at the patient area; it was highly significant when disposable clothing was used while it was weaker when cotton clothing was worn. A good correlation was also found in operating theaters supplied with unidirectional flow when disposable clothing was worn by the surgical team. This result was attributed to the increased emission by cotton clothing compared to disposable, which carries smaller bacteria particles collected from the active sampler, but are easily removed by the flow of air which is not as prone to settle on the wound, as on the plate [76,153].

The lack of correlation between active and passive sampling could be due to the punctiform active sampling compared with the longer exposure of settle plates, so that the two methods do not record omogenous situations.

This result may be due to the fact that the sedimentation plates were exposed for one hour, while the active sampling was carried out for a short period of time. Such difference, in fact, was not observed by performing the active sampling, which levies serial over-exposure of the plate [131].

Perdelli et al. evaluated the air microbial contamination in 25 operating theatres during activity by an SAS sampler and settle plates and highlighted that a significant correlation emerged only when an omogenous sampling was assured. The highest correlation was found when comparing the results obtained with passive sampling and the average of the active samples collected more than once (at least twice) during those obtained with the sampler active with samples taken every 15 minutes in the hour of exposure of the plate sedimentation [160].

In several recent studies performed in operating theaters and dental clinics, a correlation was found between the two methods [40,41,157,158]. Petti et al. found a significant correlation between the counts obtained with the SAS sampler and settle plates in a dental clinic; however for low contamination levels, this association was not significant [161].

If airborne particles are too small to settle, high values of CFUm^3 can be recorded without correspondingly high values on settle plates. Because single fungal spores can remain suspended in the air indefinitely, the use of settle plates is not recommended when sampling air for fungal spores.

Several studies found settle plates less sensitive for the collection of fungi than active samplers [41,69,75,137,166]. In particular, Verhoeff showed a strong correlation between the fungal counts obtained using settle plates and CFU/m^3 obtained by Andersen sampler -6 stages, even if the number of fungal species detected with the plates of sedimentation was significantly lower than the number recorded by using the Andersen sampler [137].

In a study by Pasquarella et al., performed in operating theaters, it was found that for empty operating theaters, the active sampler alone was able to detect the presence of fungi in 36.84% of samplings; in 7.89% of samplings fungi were also recorded by the settle plates; for working theaters the active sampler alone detected fungi in 40.81% of samplings; in 2.72% of cases, fungal contamination was only detected by the settle plates; in 13.60% of cases, both types of samplings were able to detect fungal contamination [41]. In the study performed in an operating theater by Napoli et al, only two strains of fungi were identified, one by an active sampler and one by settle plates [157]. Asefa et al. investigated performances of the SAS-super-180 air sampler and settle plates for the assessment of airborne fungal contamination in a food processing plant [69]. The performance of settle plates to capture viable fungal spores was found to be affected by the total concentration of spores in the air. At a high concentration of spores measured by the air sampler, moderately high numbers of colonies were also observed on settle plates. In sites with a low concentration of spores on SAS plates, empty settle plates or very low colony counts were obtained. This is supported by the significant correlation obtained between active and passive sampling. Active sampling allowed the recovery of a higher mean number of fungal species; 11 fungal species were recovered by the active sampling, which were not captured by settle plates, compared to the recovery of only two species by the settle plates, which were not captured by the air sampler. However, the authors underline the fact that many of the species that were recovered only by the air sampler are not known to be associated with the mycobiota of dry-cured meat products. Settle plates have the ability to provide information regarding the dominant airborne fungal spores, which can fall and cause contamination on the surface of food products.

CONCLUSION

The microbiological sampling of air is a useful tool for the collection of information aimed at understanding a phenomenon, in order to make decisions and set up targeted preventive measures. The microbiological sampling must be preceded by a careful analysis of its usefulness with a multidisciplinary approach. It is important to carefully consider the objective of the sampling and, once it has been defined, choose the most appropriate method.

As indicated by the ISO 14698-1 standard, active and passive sampling have different purposes: active sampling provides information about the concentration of viable particles in the air, whereas passive sampling provides a measure of the contribution made by aerobiocontamination to the biocontamination of surfaces. Therefore, they should not be seen as antagonists, but each of them have advantages and limits, which must be considered. Both can be used for a generic assessment of the air quality, but for specific objectives, the choice may fall on either. For example, for a monitoring carried out to estimate the risk of contamination of a critical surface, the passive sampling is the easiest and most useful method, while the evaluation of the concentration of microorganisms in the air or the estimation of biological particles that are inhaled requires the use of active sampling.

Special attention should be given to microbiological sampling procedures that include a written, well-defined protocol for sample collection. The parameter of sampling need to be carefully considered, as the standardization of the methods is the essential prerequisite for the comparison of the data. Attention should be paid to the transport of the samples and their possible conservation, the processing of the samples, the analysis and interpretation of the results and, above all, the implementation of interventions based on the results obtained.

Careful execution of air sampling by operators who are trained and motivated is essential [57].

The microbial monitoring of air should be viewed as a tool to assess the effectiveness of preventive interventions aimed at reducing microbial air contamination and identifying critical situations in the continuous process of quality improvement.

REFERENCES

[1] Al-Dagal, M; Fung, DY. Aeromicrobiology - a review. *Crit Rev Food Sci Nutr*, 1990; 29, 333-40.

[2] American Conference of Governmental Industrial Hygienists (ACGIH). Documentation of the threshold limits values and biological exposure indices. 2002.

[3] Burfoot, D. Aerosol as a contamination risk. In: Lelieveld HLM, Mostert MA, Holah J (Eds). Handbook of hygiene in the food industry. Cambridge: Woodhead Publishing Limited, 2005, 93-102.

[4] Castro, VA; Thrasher, AN; Healy, M; Ott, CM; Pierson, DL. Microbial characterization during the early habitation of the International Space Station. *Microb Ecol*, 2004; 47, 119-26.

[5] Centers for Disease Control and Prevention (CDC). Guidelines for environmental infection control in health-care facilities. 2003.

[6] Centers for Disease Control and Prevention. Guidelines for infection control in dental health-care settings. *MMWR* 2003.

[7] Centers for Disease Control and Prevention. Guidelines for preventing opportunistic infections among hematopoietic stem cell transplant recipients. 2000.

[8] Cheng, SM; Streifel, AJ. Infection control considerations during construction activities: land excavation and demolition. *Am J Infect Control*, 2001; 29, 321-8.

[9] Cobo, F; Stacey, GN; Cortés, JL; Concha, A. Environmental monitoring in stem cell banks. *Appl Microbiol Biotechnol*, 2006; 70, 651-62.

[10] Cole, EC; Cooke, CE. Characterization of infectious aerosols in health care facilities: an aid to effective engineering controls and preventive strategies. *Am J Infect Control*, 1998; 26, 453-64.

[11] Conseil Supérieur de la Santé. Recommandations en matière de contrôles bactériologiques de l'environnement dans les institutions de soins. 2010.

[12] Decraene, V; Ready, D; Pratten, J; Wilson, M. Air-borne microbial contamination of surface in a UK dental clinic. *J Gen Appl Microbiol*, 2008; 54, 195-203.

[13] Dharan, S; Pittet, D. Environmental control in operating theatres. *J Hosp Infect*, 2002; 51, 79-84.

[14] Eickhoff, TC. Airborne nosocomial infection: a contemporary perspective. *Infect Control Hosp Epidemiol*, 1994; 15, 663-72.

[15] European Collaborative Action. Indoor air quality & its impact on man. Report N. 12. Biological particles in indoor environments. Commission of the European Communities. EUR 14988 EN 1993.

[16] European Commision. EC Guide to good manufacturing practice. Revision to Annex 1. Manufacture of Sterile Medicinal Products. Brussels, Belgium, 2008.

[17] Ferretti, S; Pasquarella, C; Fornia, S; et al. Effect of a mobile unidirectional air flow unit on microbial contamination of air in standard urologic procedures. *Surg Infect*, 2009; 10, 511-6.

[18] Guarnieri, V; Gaia, E; Battocchio, L; et al. New methods for microbial contamination monitoring: an experiment on board the MIR orbital station. *Acta Astronautica*, 1997; 40, 195-201.

[19] Health Technical Memorandum 03-01. Specialised ventilation for healthcare premises-Part A: design and validation; 2007.

[20] Hofstra, H; van der Vossen, JMBM; van der Plas, J. Microbes in food processing technology. *FEMS Microbiol Rev*, 1994; 15, 175-83.

[21] Hryhorczuk, D; Curtis, L; Scheff, P; et al. Bioaerosol emissions from a suburban yard waste composting facility. *Ann Agric Environ Med*, 2001; 8, 177-185.

[22] Istituto Nazionale per l'Assicurazione contro gli Infortuni sul Lavoro (INAIL). Consulenza Tecnica Accertamento Rischi. Il monitoraggio microbiologico negli ambienti di lavoro. Campionamento e analisi. 2010.

[23] Istituto Superiore per la Prevenzione e la Sicurezza del Lavoro (ISPESL). Linee guida per la definizione degli standard di sicurezza e di igiene ambientale dei reparti operatori. 2009.

[24] Jensen, PA; Schafer, MP; NIOSH/DPSE. Sampling and characterization of bioaerosols. *NIOSH Manual Analytical Methods*. 1998.

[25] Kang, YJ; Frank, JF. Biological aerosol: a review of airborne contamination and its measurement in dairy processing plants. *J Food Protect*, 1989; 52, 512-24.

[26] Krajewski, JA; Tarkowski, S; Cyprowski, M; Szarapinska-Kwaszewska, J; Dudkiewicz, B. Occupational exposure to organic dust associated with municipal waste collection and management. *Int J Occup Med Environ Health*, 2002; 15, 289-301.

[27] Laumbach, RJ; Kipen, HM. Bioaerosol and sick building syndrome; particles, inflammation and allergy. *Curr Opin Allerg Clin Immunol*, 2005; 5, 135-9.

[28] Lee, SA; Adhikari, A; Grinshpun, SA; Mckay, R; Shukla, R; Reponen, T. Personal exposure to airborne dust and microorganisms in agricultural environments. *J Occup Environ Hyg*, 2006; 3, 118-30.

[29] Lindquist, TD; Miller, TD; Elsen, JL; Lignoski, PJ. Minimizing the Risk of Disease Transmission During Corneal Tissue Processing. *Cornea*, 2009; 28, 481-4.

[30] Lidwell, OM. The microbiology of air. In: Topley & Wilson's. Principles of bacteriology, virology and immunity. *General bacteriology and immunity*. Vol. 1. London: Edward Arnold, 1990, 226-39.

[31] Loddon Mallee Region, Infection Control Resource Center. Infection Control Principles for the Management of Construction, Renovation, Repairs and Maintenance within Health Care Facilities. A Manual for Reducing the risk of Health Care Associated Infection by Dust and Water Borne Microorganisms. 2nd Edition. 2003; Reviewed August 2005.

[32] Mandrioli, P; Caneva, G. Cultural heritage and aerobiology. *Methods and measurement techniques for Biodeterioration Monitoring.* Dordrecht: Springer, 2004.

[33] NHS Estates. Health Technical Memorandum 2025. *Ventilation in Healthcare Premises*. Management Policy 1994.

[34] Nusca, A; Bonadonna, L; Orefice, L. Diffusion of biological agents in air in a closed environment and related pathologies. *Ig San Pubbl*, 2003; 59, 175-86.

[35] Pasqualotto, AC; Denning, DW. Post-operative aspergillosis. *Clin Microbiol Infect*, 2006; 12, 1060-76.

[36] Pasquarella, C; Agodi, A. Infezioni di origine aerea nelle strutture sanitarie. *Ann Ig*, 2007; 19 (Suppl. 1), 29-44.

[37] Pasquarella, C; Pitzurra, O; Herren, T; Poletti, L; Savino, A. Lack of influence of body exhaust gowns on aerobic bacterial surface counts in a mixed-ventilation operating theatre. A study of 62 hip arthroplaties. *J Hosp Infect*, 2003; 54, 2-9.

[38] Pasquarella, C; Saccani, E; Sansebastiano, GE; Ugolotti, M; Pasquariello, G; Albertini, R. Proposal for a biological environmental monitoring approach to be used in libraries and archives. *Ann Agric Environ Med*, 2012; 19, 201-4.

[39] Pasquarella, C; Sansebastiano, GE; Saccani, E; et al. Proposal for an integrated approach to microbial environmental monitoring in cultural heritage: experience at the Correggio exhibition in Parma. *Aerobiologia*, 2011; 27, 203-211.

[40] Pasquarella, C; Veronesi, L; Napoli, C; et al. Microbial environmental contamination in Italian dental clinics: A multicenter study yielding recommendations for standardized sampling methods and threshold values. *Sci Total Environ*, 2012; 420, 289-99.

[41] Pasquarella, C; Vitali, P; Saccani, E; et al. Microbial air monitoring in operating theatres: experience at the University Hospital of Parma. *J Hosp Infect*, 2012; 81, 50-57.

[42] Pasquarella, C. Microbial control of the environment in the operating theatre. *Ann Ig*, 2009; 21(Suppl 1).

[43] Peláez, T; Muñoz, P; Guinea, J; Giannella, M; Klaassen, CH; Bouza, E. Outbreak of invasive aspergillosis after major heart surgery caused by spores in the air of the intensive care unit. *Clin Infect Dis*, 2012; 54, 24-31.

[44] Perdelli, F; Sartini, M; Spagnolo, AM; Dallera, M; Lombardi, R; Cristina, ML. A problem of hospital hygiene: the presence of aspergilli

in hospital wards with different air-conditioning features. *Am J Infect Control*, 2006; 34, 264-8.

[45] Rautemaa, R; Nordberg, A; Wuolijoki-Saaristo, K; Meurman, JH. Bacterial aerosol in dental practice - a potential hospital infection problem? *J Hosp Infect*, 2006; 64, 76-81.

[46] Rautiala, S; Kangas, J; Louhelainen, K; Reiman, M. Farmers' exposure to airborne microorganisms in composting swine confinement buildings. *AIHA J*, 2003; 64, 673-7.

[47] République Francaise, Ministere de la Santé, del la famille et des persone handicapees. Surveillance microbiologique de l'environnement dans les établissements de santé. Air, eaux et surfaces. 2002.

[48] Rolka, H; Krajewska-Kułak, E; Łukaszuk, C; et al. Indoor air studies of fungi contamination of social welfare home in Czerewki in north-east part of Poland. Annales Academiae Medicae Bialostocensis. *Annual Proceedings of Medical Science,* 2005;50, 26-30.

[49] Salik, K; Fleischer, M. Postepowanie w przypadku wystapiena szpitalnych ognisk epidemicznych. 2006.

[50] Scherwing, C; Golin, F; Guenec, O; et al. Continuous microbiological air monitoring for aseptic filling lines. *PDA J Pharm Scien Technol*, 2007; 61, 102-9.

[51] Schillinger, JE; Vu, T; Bellin, P. Airborne fungi and bacteria: background levels in office buildings. *J Environm Health*, 1999; 62, 9-14.

[52] Société francaise d'Hygiéne Hospitaliére. Risque infectieux fongique et travaux en établissement de santé. Identification du risque et mise en place de mesures de gestion. Vol XIX, N° 1, 2011.

[53] Société francaise d'Hygiéne Hospitaliére. La qualité de l'air au bloc opératoire. Recommandations d'Experts. GR-AIR/Octobre 2004.

[54] Thorn, J; Kerekes, E. Health effects among employees in sewage treatment plants: a literature survey. *Am J Ind Med*, 2001; 40, 170-9.

[55] U.S. Pharmacopeia 30 - National Formulary 25, <1116>Microbiological evaluation of clean rooms and other controlled environments.

[56] Vonberg, RP; Gastmeier, P. Nosocomial aspergillosis in outbreak settings. *J Hosp Infect*, 2006; 63, 246-54.

[57] UNI EN 13098:2002. Atmosfera nell'ambiente di lavoro - Linee guida per la misurazione di microrganismi e di endotossine aerodispersi.

[58] UNI EN ISO 14698-1: 2004. Camere bianche ed ambienti associati controllati - Controllo della biocontaminazione - Parte 1: Principi generali e metodi.

[59] Utrup, LJ; Frey, AH. Fate of bioterrorism-relevant viruses and bacteria, including spores, aerosolized into an indoor air environment. *Exp Biol Med*, 2004; 229, 345-50.

[60] UNI EN ISO 14644-1: 2001 Camere bianche ed ambienti associati controllati - Classificazione della pulizia dell'aria.

[61] Möller, AL. Measurement of airborne microorganisms - a long development still with problems. *Monitor*, 2000; 45, 2-12.

[62] Pasteur, L. Mémoire sur le corpuscles organisés qui existent dans l'atmosphère: examen de la doctrine de générations spontanées. *Ann Sci Nat*, 186; 16, 5-98.

[63] Tyndall, J. *Floating-matter of air*. New York: Appleton. 1882.

[64] Kock, R. Zur Untersuchung von pathogenen Organismen, Mitt. A.d.k. Gsndhsamte, Berlin, 1, 1-48, 234-282. 1881.

[65] Wells, WF. Apparatus for study of the bacterial behavior of air. *Am J Pub Health*, 1933; 23, 58-59.

[66] Bourdillon, RB; Lidwell, OM; Thomas, JC. A slit sampler for collecting and counting air-borne bacteria. *J Hyg*, 1941; 41, 198-224.

[67] Lidwell, OM. Sir John Charnley, Surgeon (1911-82): the control of infection after total joint replacement. *J Hosp Infect*, 1993; 23, 5-15.

[68] Andon, BM. Active air vs. passive air (Settle plates) monitoring in routine environmental monitoring programs. *PDA Journal of Pharmaceutical Science and Techonology*, 2006; 60, 350-5.

[69] Asefa, DT; Langsrud, S; Gjerde, RO; Kure, CF; Sidhu, MS; Nesbakken, T; Skaar, I. The performance of SAS-super-180 air sampler and settle plates for assessing viable fungal particles in the air of dry-cured meat production facility. *Food Control*, 2009; 20, 997-1001.

[70] Blomquist, G. Sampling of biological particles. *Analyst*, 1994; 119, 53-6.

[71] Burge, HA; Solomon, WR. Sampling and analysis biological aerosols. *Atmos Environ*, 1987; 21, 451-6.

[72] Buttner, MP; Willeke, K; Grinshpun, S. Sampling and analysis of airborne microorganisms. In: Hurst CJ, Knudsen GR, Mcclnerney MJ, Stetzenbach LD, Walzer MD. *Manual of Environmental Microbiology*. Washington DC: American Society for Microbiology 1997; 629-40.

[73] Charnley, J. Post operative infection after total hip replacement with special reference to air contamination in the operating room. *Clin Orthop*, 1972; 87, 167-87.

[74] Costa Salustiano, V; Andrade, NJ; Cardoso Brandao, SC; Cordeiro Azeredo, RM; Kitakawa Lima, SA. Microbiological air quality of

processing areas in a diary plants as evaluated by the sedimentation technique and one-stage air sampler. *Brazilian Journal of Microbiology*, 2003; 34, 255-9.

[75] French, MLV; Eitzen, HE; Ritter, MA; Leland, DS. Environmental control of microbial contamination in the operating room. In: Hunt TK, Ed. Wound Healing and Wound Infection. New York: Appleton-Century Crofts, 1980; 254-61.

[76] Friberg, B; Friberg, S; Burman, LG. Inconsistent correlation between aerobic bacterial surface and air counts in operating rooms with ultraclean laminar air flows: proposal of a new bacteriological standard for surface contamination. *J Hosp Infect*, 1999; 42, 287-93.

[77] Griffiths, WD; DeCosemo, GAL. The assessment of bioaerosols: A critical review. *J Aerosol Sci*, 1994; 25, 1425-58.

[78] Grillot, R; Nolard, N. Surveillance de l'environnement des malades à risque fongique: méthodes d'évaluation et utilité. *Hygienes*, 2000; 8, 408-17.

[79] Gröschel, DHM. Air sampling in hospitals. *Annals New York Academy Sciences*, 1980; 353, 230-9.

[80] Hoffman, P; Humphreys, H. Air sampling: settle plates or slit samplers? *J Hosp Infect*, 2001; 299-300.

[81] Kundsin, RB. The microbiologist's role in evaluating the hygienic environment. In: Kundsin RB ed. Architectural design and indoor microbial pollution. Oxford University Press, 1988, 103-21.

[82] Muñoz, P; Burillo, A; Bouza, E. Environmental surveillance and other control measures in the prevention of nosocomial fungal infections. *Clin Microbiol Infect*, 2001; 7, 38-45.

[83] Napoli, C. Prevention of Healthcare-Associated Infections: Which Sampling Method should be used to evaluate Air Bio-Contamination in Operating Rooms? *Epidemiol*, 2012; 2, 2.

[84] Nevalainen, A; Pastuszka, J; Liebhaber, F; Willeke, K. Performance of bioaerosol samplers: collection characteristics and sampler design considerations. *Atmospheric Environment*, 1992; 26, 531-40.

[85] Pasquarella, C; Pitzurra, O; Savino, A. The index of microbial air contamination. *J Hosp Infect*, 2000; 46, 241-56.

[86] Whyte, W. In support of settle plates. *PDA J Pharm Scien Technol*, 1996; 50, 201-4.

[87] Davies, RR; Noble, WC. Dispersal of bacteria in desquamated skin. *Lancet*, 1962; 2, 1295-7.

[88] Whyte, W; Hejab, M. Particle and microbial airborne dispersion from people. *European Journal of Parenteral & Pharmaceutical Sciences*, 2007; 12, 39-46.

[89] Wells, WF. On air-borne infection. Study II. Droplets and droplet nuclei. *Am J Hyg*, 1934; 20, 611-8.

[90] Greene, VW; Vesley, D; Bond, RG; Michaelsen, GS. Microbiological contamination of hospital air. II. Quantitative studies. *Lancet*, 1962; 10, 561-6.

[91] Greene, VW; Vesley, D; Bond, RG; Michaelsen, GS. Microbiological contamination of hospital air. II. Qualitative studies. *Lancet*, 1962; 10, 567-71.

[92] Andersen, AA. New sampler for the collection, sizing and enumeration of viable airborne particles. *J Bacteriol*, 1958; 76, 471-84.

[93] Chang, C; Hwang, YH; Grinshpun, SA; Macher, JM; Willeke, K. Evaluation of counting error due to colony masking in bioaerosol sampling. *Appl Environ Microbiol*, 1994; 60, 3732-3738.

[94] Cage, BR; Schreiber, K; Barnes, C; Portnoy, J. Evaluation of four bioaerosol samplers in the outdoor environment. *Ann Allergy Asthma Immunol*, 1996; 77, 401-6.

[95] Whyte, W; Green, G; Abisu, A. Collection efficiency and design of microbial samplers. *Aerosol Science*, 2007; 38, 101-14.

[96] Stewart, SL; Grinshpun, SA; Willeke, K; Terzieva, S; Ulevicius, V; Donnelly, J. Effect of impact stress on microbial recovery on an agar surface. *Appl Environ Microbiol*, 1995; 1232-9.

[97] An, HR; Mainelis, G; Yao, M. Evaluation of high - volume portable bioaerosol sampler in laboratory and field environments. *Indoor Air*, 2004; 14, 385-93.

[98] Lach, V. Performance of the surface air system air samplers. *J Hosp Infect*, 1985; 6, 102-7.

[99] Macher, JM; First, MW. Reuter centrifugal air sampler: measurement of effective air flow rate and collection efficiency. *Appl Environ Microbiol*, 1983; 45, 1960-2.

[100] Jensen, PA; Todd, W; Davis, GN; Scarpino, P. Evaluation of eight bioaerosol samplers challenged with aerosols of free bacteria. *Am Ind Hyg Assoc J,* 1992; 53, 660-7.

[101] Parks, SR; Bennett, AM; Speight, SE; Benbough, JE. An assessment of the Sartorius MD8 microbiological air sampler. *J Appl Bacteriol*, 1996; 80, 529-34.

[102] May, KR; Harper, GJ. The efficiency of various liquid impinger samplers in bacterial aerosols. *Br J Ind Med*, 1957; 14, 287-97.

[103] Chang, C; Hwang, Y; Grinshpun, SA; Macher, JM; Willeke, K. Evaluation of counting error due to colony masking in bioaerosol samping., 1994; 60, 3732-8.

[104] Buttner, MP; Setzenbah, LD. Evaluation of four aerobiological sampling methods for the retrieval of aerosolized Pseudomonas syringae. *Appl Environm Microbiol*, 1991; 57, 1268-70.

[105] Godish, DR; Godish, TJ. Relationship between sampling duration and concentration of cultural airborne and bacteria on selected culture media. *J Appl Microbiol*, 2007; 102, 1479-84.

[106] Pongiluppi, S. Esperienze di controllo microbiologico e criteri di accettabilità per le produzioni in blocco sterile. In: Atti "Controllo microbiologico dell'aria e dell'igiene ambientale". Milano, 2-4 marzo 1983.

[107] Whyte, W; Lidwell, OM; Lowbury, EJL; Blowers, R. Suggested bacteriological standards for air in ultra-clean operating rooms. *J Hosp Infect*, 1983; 4, 133-9.

[108] Hoffman, PN; Williams, J; Stacey, A; et al. Microbiological commissioning and monitoring of operating suites. *J Hosp Infect*, 2002; 52, 1-28.

[109] Thio, CL; Smith, D; Merz, WG; et al. Refinements of environmental assessment during an out break investigation of invasive aspergillosis in a leukaemia and bone marrow transplant unit. *Inf Control Hosp Epidemiol*, 2000; 21, 18-23.

[110] Streifel, AJ. Aspergillus and construction. In: Kundsin RB, ed. *Architectural Design and Indoor Microbial Pollution*. New York, NY: Oxford University Press; 1988, 198-216.

[111] Ambroise, D; Barbotte, E; Hartemann, P. Comparing microbial performance of impactor air samplers: an uneasy task. *J Hosp Infect*, 2002; 51, 145-6.

[112] Ambroise, D; Greff-Mirguet, G; Görner, P; Fabriès, JF; Hartemann, P. Measurement of indoor viable airborne bacteria with different bioaerosol samplers. *J Aerosol Sci*, 1999; 30 (Suppl. 1), 699-700.

[113] Bellin, P; Schillinger, J. Comparison of field performance of the Andersen N6 single stage and the SAS sampler for airborne fungal propagules. *Indoor Air*, 2001; 11, 65-8.

[114] Bonadonna, L; Marconi, A. A comparison of two air samplers for recovery of indoor bioaerosols. *Aerobiologia*, 1994; 10, 153-6.

[115] Buttner, MP; Stetzenbach, LD. Monitoring airborne fungal spores in an experimental indoor environment to evaluate sampling methods and the effects of human activity on air sampling. *App Environ Microbiol*, 1993; 219-26.

[116] Casewell, MW; Fermie, PG; Thomas, C; Simmons, NA. Bacterial air counts obtained with a centrifugal (RCS) sampler and a slit sampler - the influence of aerosols. *J Hosp Infect*, 1984; 5, 76-82.

[117] Clark, S; Lach, V; Lidwell, OM. The performance of the Biotes RCS centrifugal air sampler. *J Hosp Infect*, 1981; 2, 181-6.

[118] Curtis, SE; Balsbaugh, RK; Drummond, JG. Comparison of Andersen eight-stage and two-stage viable air sampler. *Appl Environ Microbiol*, 1978; 35, 208-9.

[119] Delmore, RP; Thompson, WN. A comparison of air-sampler efficiency. *Med Device Diagn Ind*, 1981; 3, 45-8.

[120] Gangneux, JP; Robert-Gangeneux, F; Gicquel, G; et al. Bacterial and fungal counts in hospital air: comparative yields for 4 sieve impactor air samplers with 2 culture media. Infect *Control Hosp Epidemiol*, 2006; 27, 1405-8.

[121] Hartemann, P; Blech, MF; Ruden, H. Etude comparative de plusieurs appareils de mesure de la biocontamination de l'air. *VDI-Berichte* Nr 386, 1980.

[122] Kang, YJ; Frank, JF. Evaluation of air samplers for recovery of biological aerosols in dairy processing plants. *J Food Protect*, 1989; 52, 655-9.

[123] Lee, KS; Bartlett, KH; Brauer, M; Stephens, GM; Black, WA; Teschke, K. A field comparison of four samplers for enumerating fungal aerosols. I. Sampling characteristics. *Indoor Air*, 2004; 14, 360-6.

[124] Lembke, LL; Kinseley, RN; Nostarnd, RCV; Hale, MD. Precision of the all glass impinger and the Andersen microbial impactor for air sampling in a solid-waste handling facilities. *Appl Environm Microbiol*, 1981; 42, 222-5.

[125] Liguori, G; Spagnoli, G; Agozzino, E; et al. Biological risk in operating room: microbiological monitoring of the environment and analysis of the associated variables. *An Ig*, 2005; 17, 385-400.

[126] Ljungqvist, B; Reinmuller, B. Active sampling of airborne viable particles in controlled environments: a comparative study of common instruments. *E J Par Scien*, 1998; 3, 59-62.

[127] Lundholm, IM. Comparison of methods for quantitative determinations of airborne bacteria and evaluation of total viable counts. *Appl Environ Microbiol*, 1982; 44, 179-83.

[128] Mehta, SK; Mishra, SK; Pierson, DL. Evaluation of three portable samplers for monitoring airborne fungi. *App Environ Microbiol*, 1996; 62, 1835-8.

[129] Morris, G; Kokki, MH; Anderson, K; Richardson, MD. Sampling of Aspergillus spores. *J Hosp Infect*, 2000; 44, 81-92.

[130] Nakhla, LS; Cummings, RF. A comparative evaluation of a new centrifugal air sampler (RCS) with a slit air sampler (SAS) in a hospital environment. *J Hosp Infect*, 1981; 2, 261-6.

[131] Nesa, D; Lortholary, J; Boukline, A; et al. Comparative performance of impactor air samplers for quantification of fungal contamination. *J Hosp Infect*, 2001; 47, 149-55.

[132] Pitzurra, M; Pasquarella, C; Pitzurra, O; Savino, A. The measure of microbial contamination of the air. Part I. *Ann Ig*, 1996; 8, 349-59.

[133] Placencia, AM; Peeler, JT; Oxborrow, GS; Danielson, JW. Comparison of bacterial recovery by Reuter Centrifugal Air Sampler and Slit-to-Agar Sampler. *Appl Environm Microbiol*, 1982; 44, 512-3.

[134] Shintani, H; Taniai, E; Miki, A; Kurosu, S; Hayashi, F. Comparison of the collecting efficiency of microbiological air samplers. *J Hosp Infect*, 2004; 56, 42-8.

[135] Upton, SL; Mark, D; Douglass, EJ; Hall, DJ; Griffiths, WD. A wind tunnel evaluation of the physical sampling efficiency of three bioaerosol samplers. *J Aerosol Science*, 1994; 25, 1495-501.

[136] Verhoeff, A; van Wijnen, J; Attwood, P; Bolei, J; Brunekreef, B; van Reenen-Hoekstra, ES; Samson, R. Enumeration and identification of airborne viable mould propagules in houses: a comparison of selected measurement techniques University of Wageningen, Holland, 1988.

[137] Verhoeff, AP; van Wijnen, JH; Boleij, JSM; Brunekreef, B; van Reenen-Hoekstra, ES; Samson, RA. Enumeration and identification of airborne viable mould propagules in houses. *Allergy*, 1990; 45, 275-84.

[138] Wu, PC; Su, HJ; Ho, HM. A comparison of sampling media for environmental viable fungi collected in a hospital environment. *Environ Res*, 2000; 82, 253-7.

[139] Verhoeff, AP; van Wijnen, JH; Brunekreef, B; Fisher, P; van Reenen-Hoekstra, ES; Samson, RA. Presence of viable mould propagules in indoor air in relation to house damp and outdoor air. *Allergy*, 1992; 47, 83-91.

[140] Whyte, W. The Casella Slit sampler or the Biotest Centrifugal sampler - which is more efficient? *J Hosp Infect*, 1981; 2, 297-9.

[141] Tavora, LGF; Gambale, W; Heins-Vaccari, EM; et al. Comparative performance of two air samplers for monitoring airborne fungal propagules. *Braz J Med Biol Res*, 2003; 36, 613-6.

[142] Tolchinsky, AD; Sigaev, VI; Varfolomeev, AN; Uspenskaya, SN; Cheng, YS; Su, WC. Performance evaluation of two personal bioaerosol samplers. *Journal Environmental Science and Health,* Part A, 2011; 46, 1690-1698.

[143] Xu, Z; Wei, K; Wu, Y; Shen, F; Chen, Q; Li, M; Yao, M. Enhancing bioaerosol sampling by Andersen impactors using mineral-oil-spread agar plate. *Plose One*, 2013; 8, 1-10.

[144] Cooper, EE; O'Reilly, MA; Guest, DI; Dharmage, SC. Influence of building construction work on *Aspergillus* infection in a hospital setting. *Infect Control Hosp Epidemiol*, 2003; 24, 472-6.

[145] Regione Lombardia. *Linee guida sulla prevenzione e sicurezza nelle sale operatorie. Dicembre 1999.* Approvate con DGR 17 dicembre 1999 n. 6/47077. Bol Uff Reg Lombardia 4 gennaio 2000, 1° Suppl Straordinario.

[146] Pitzurra, M; D'Alessandro, D; Pasquarella, C; et al. Indagine su caratteristiche e modalità di gestione degli impianti di condizionamento dell'aria in alcune sale operatorie italiane. *Ann Ig*, 1997; 9, 429-38.

[147] Mammarella, L. Inquinamenti dell'aria. Roma: Il Pensiero Scientifico Editore, 1971, 30-52.

[148] Fisher, G; Fodré, S; Nehez, M. Neuere Beitrage zur Standardisierung mit mikrobiologischen Sedimentations Luftuntersuchungen. *Z Ges Hyg*, 1972; 18, 267-71.

[149] Dancer, SJ. How do we assess hospital cleaning? A proposal for microbiological standards for surface hygiene in hospitals. *J Hosp Infect*, 2004; 56, 10-5.

[150] Whyte, W; Hodgson, R; Tinkler, J. The importance of airborne bacterial contamination of wounds. *J Hosp Infect*, 1982; 3, 123-35.

[151] Humphreys, H; Stacey, AR; Taylor, EW. Survey of operating theatres in Great Britain and Ireland. *J Hosp Infect*, 1995; 30, 245-52.

[152] Holton, J; Ridgway, GL. Commissioning operating theatres. *J Hosp Infect*, 1993; 23, 153-60.

[153] Friberg, B; Friberg, S; Burman, LG. Correlation between surface and air counts of particles carrying aerobic bacteria in operating rooms with

turbulent ventilation: an experimental study. *J Hosp Infect*, 1999; 42, 61-8.

[154] Russell, MP; Goldsmith, JA; Philips, I. Some factors affecting the efficiency of settle plates. *J Hosp Infect*, 1984; 5, 189-99.

[155] Whyte, W; Niven, L. Airborne bacteria sampling: the effect of dehydration and sampling time. *J Parent Sci Technol*, 1986; 40, 182-7.

[156] Fiorina, A; Olivo, F; Pensi, N; Carcheri, G; Macrina, G; Crimi, P. A portable sampler (Partrap FA 52) for microbiological evaluation of airborne particles: comparison with standard sedimetric and volumetric methods in haemodialysis rooms. *Ann Microbiol*, 2006; 56, 283-7.

[157] Napoli, C; Marcotrigiano, V; Montagna, MT. Air sampling procedures to evaluate microbial contamination: a comparison between active and passive methods in operating theatres. *BMC Public Health*, 2012; 12, 594.

[158] Napoli, C; Tafuri, S; Montenegro, L; Cassano, M; Notarnicola, A; Lattarulo, S; Montagna, MT; Moretti, B. Air sampling methods to evaluate microbial contamination in operating theatres: results of a comparative study in an orthopaedics department. *J Hosp Infect*, 2012; 80, 128-132.

[159] Orpianesi, C; Cresci, A; La Rosa, F; Saltalamacchia, G; Tarsi, R. Evaluation of microbial contamination in a hospital environment. Comparison between the Surface Air System and the traditional method. *Ann Ig*, 1983; 34, 171-85.

[160] Perdelli, F; Sartini, M; Orlando, M; Secchi, V; Cristina, ML. Relationship between settling microbial load and suspended microbial load in operating rooms. *Ann Ig*, 2000; 12, 373-80.

[161] Petti, S; Iannazzo, S; Tarsitani, G. Comparison between different methods to monitor the microbial level of indoor air contamination in the dental office. *Ann Ig*, 2003; 15, 725-33.

[162] Pitzurra, M; Morlunghi, P. Contaminazione microbica dell'aria atmosferca: correlazione tra due diverse metodiche di rilevazione. *Ig Mod*, 1978; 3, 489-501.

[163] Sayer, WJ; MacKnight, NM; Wilson, HW. Hospital airborne bacteria as estimated by the Andersen sampler versus gravity settling culture plate. *Am J Clin Pathol*, 1972; 58, 558-62.

[164] Pasquarella, C; Masia, MD; Nnanga, N; et al. Microbial air monitoring in operating theatre: active and passive samplings. *Ann Ig*, 2004; 16, 375-86.

[165] Albertini, R; Pasquarella, C; Sansebastiano, GE; et al. Esperienze di monitoraggio aerobiologico presso l'Azienda Ospedaliero-Universitaria di Parma. In: Atti XI Congresso Nazionale di Aerobiologia. Parma, 5-8 aprile 2006. *Giornale Europeo di Aerobiologia*. Suppl 1/2006, 47-8.
[166] Solomon, WR. Assessing fungus prevalence in domestic interiors. *Journal of Allergy and Clinical Immunology*, 1975; 56, 235-42.

In: Air Quality
Editor: Arthur Hermans

ISBN: 978-1-62808-259-3

Chapter 2

APPLICATION OF A MARS-BASED REGRESSION MODEL TO THE AIR QUALITY STUDY AT LOCAL SCALE IN THE GIJÓN URBAN AREA (NORTHERN SPAIN)

P. J. García Nieto[1,*] and J. C. Álvarez Antón[2]

[1]Department of Mathematics, University of Oviedo, Oviedo, Spain
[2]Department of Electrical Engineering, Campus de Viesques, University of Oviedo, Gijón, Spain

ABSTRACT

The aim of this research work is to build a regression model of air quality by using the multivariate adaptive regression splines (MARS) technique in the Gijón urban area (Northern Spain) at local scale. With the rapid increases in processing speed and memory of low-cost computers, it is not surprising that various advanced computational learning tools such as neural networks have been increasingly used for analyzing or modeling highly nonlinear multivariate engineering problems. However, neural networks have been criticized for its long training process since the optimal configuration is not known a priori. This research work explores the use of a nonparametric regression algorithm known as multivariate adaptive regression splines (MARS)

*Corresponding author. Tel.: +34-985103417; Fax: +34-985103354
E-mail address: lato@orion.ciencias.uniovi.es (P.J. García Nieto).

which has the ability to approximate the relationship between the inputs and outputs, and express the relationship mathematically. In this sense, hazardous air pollutants or toxic air contaminants refer to any substance that may cause or contribute to an increase in mortality or serious illness, or that may pose a present or potential hazard to human health. To accomplish the objective of this study, the experimental dataset of nitrogen oxides (NO_x), carbon monoxide (CO), sulphur dioxide (SO_2), ozone (O_3) and dust (PM_{10}) were collected over 3 years (2006–2008) and they are used to create a highly nonlinear model of the air quality in the Gijón urban nucleus (Spain) based on the MARS technique. One objective of this model is to obtain a preliminary estimate of the dependence between primary and secondary pollutants in the Gijón urban area at local scale. A second aim is to determine the factors with the greatest bearing on air quality with a view to proposing health and lifestyle improvements. The United States National Ambient Air Quality Standards (NAAQS) establishes the limit values of the main pollutants in the atmosphere in order to ensure the health of healthy people. They are known as criteria pollutants. Firstly, this MARS regression model captures the main insight of statistical learning theory in order to obtain a good prediction of the dependence among the main pollutants in the Gijón urban area. Secondly, the main advantages of MARS are its capacity to produce simple, easy-to-interpret models, its ability to estimate the contributions of the input variables, and its computational efficiency. Finally, on the basis of these numerical calculations, using the multivariate adaptive regression splines (MARS) technique, conclusions of this work are exposed.

Keywords: Air quality; Pollutant substances; Statistical learning; Multivariate adaptive regression splines (MARS)

1. INTRODUCTION

Air pollution is the introduction into the atmosphere of chemicals, particulates, or biological materials that cause discomfort, disease, or death to humans, damage other living organisms such as food crops, or damage the natural environment or built environment [1-3]. The atmosphere is a complex dynamic natural gaseous system that is essential to support life on planet Earth. For instance, the stratospheric ozone depletion due to air pollution has long been recognized as a threat to human health as well as to the Earth's ecosystems. Additionally, the urban air quality are listed as one of the World's Worst Toxic Pollution Problems in the 2008 Blacksmith Institute World's

Worst Polluted Places report [4,5]. Indeed, air pollution is one of the important environmental problems in metropolitan and industrial cities [1-3] such as Gijón (Principality of Asturias, Spain). The World Health Organization states that 2.4 million people die each year from causes directly attributable to air pollution, with 1.5 million of these deaths attributable to indoor air pollution. The health effects caused by air pollution may include difficulty in breathing, wheezing, coughing and aggravation of existing respiratory and cardiac conditions. These effects can result in increased medication use, increased doctor or emergency room visits, more hospital admissions and premature death [3-5]. The human health effects of poor air quality are far reaching, but principally affect the body's respiratory system and the cardiovascular system [1]. Individual reactions to air pollutants depend on the type of pollutant a person is exposed to, the degree of exposure, the individual's health status and genetics. A new economic study of the health impacts and associated costs of air pollution in the Los Angeles Basin and San Joaquin Valley of Southern California shows that more than 3,800 people die prematurely (approximately 14 years earlier than normal) each year because air pollution levels violate federal standards. The number of annual premature deaths is considerably higher than the fatalities related to auto collisions in the same area, which average fewer than 2,000 per year. Diesel exhaust (DE) is a major contributor to combustion derived particulate matter air pollution. In several human experimental studies, using a well validated exposure chamber setup, DE has been linked to acute vascular dysfunction and increased thrombus formation. This serves as a plausible mechanistic link between the previously described association between particulate matter air pollution and increased cardiovascular morbidity and mortality [1-3,6].

Gijón is a coastal industrial city and a municipality in the autonomous community of Asturias in Spain. It was an important regional Roman city, although the area has been settled since earliest history. The main port, one of the largest in the north of Spain, is called El Musel. Gijón has a growing population of approximately 300,000 inhabitants and has an Atlantic humid maritime climate, with mild temperatures and high humidity throughout the year. The summers are warm, with both overcast and sunny days and winters are cold with significant rains. With respect to its economy, the port is at the centre of many of the local businesses. Apart from directly port related activities, the economy is based on steel (e.g. Arcelor Mittal Ltd.), other metallurgy located in industrial parks or estates (Somonte industrial park, Roces industrial park, etc.), the nearby presence of Aboño Coal-fired Power Plant, livestock rearing and fisheries, etc. Figure 1 shows the geographical location of the four meteorological stations and the most important industries in Gijón urban area including its industrial harbour: El Musel.

To fix ideas, there are many air pollution indicators affecting human health [7-11]. The automatic measurements of the meteorological pollution, such as CO, NO, NO_2, SO_2, O_3 and particulate matter (PM_{10}) is more and more important due to their harmful effects on human health [1-3]. EU and many national environmental agencies have set standards and air quality guidelines for allowable levels of these pollutants in the air [5,11,12]. The aim of this research work is to construct a model for the averaged next month pollution that would be applicable for the use by the authority responsible for air pollution regulation in the appropriate region of the country. The use of the artificial neural networks of multilayer perceptron (MLP) type as the model of pollution was exploited frequently in the last years [11,13-15]. An artificial neural network (ANN), usually called neural network (NN), is a mathematical model or computational model that is inspired by the structure and/or functional aspects of biological neural networks. A neural network consists of an interconnected group of artificial neurons, and it processes information using a connectionist approach to computation [16]. In most cases an ANN is an adaptive system that changes its structure based on external or internal information that flows through the network during the learning phase. Modern neural networks are non-linear statistical data modeling tools [17-19]. They are usually used to model complex relationships between inputs and outputs or to find patterns in data. A multilayer perceptron (MLP) is a feedforward artificial neural network model that maps sets of input data onto a set of appropriate output. A MLP consists of multiple layers of nodes in a directed graph, which is fully connected from one layer to the next. Except for the input nodes, each node is a neuron (or processing element) with a nonlinear activation function. MLP utilizes a supervised learning technique called backpropagation for training the network. MLP is a modification of the standard linear perceptron, which can distinguish data that is not linearly separable [11,14,20].

In this innovative research work, it is proposed a model based on the multivariate adaptive regression splines (MARS) [21-28] for the study of considered pollutants: CO, NO, NO_2, SO_2, O_3 and particulate matter (PM_{10}). The data taking part in learning and testing have been collected within three years: from 2006 to 2008. The results of numerical experiments based on the application of the MARS technique have confirmed good accuracy of daily modelling for all considered pollutants. These detailed results will be presented and discussed in this paper.

Indeed, the multivariate adaptive regression splines (MARS) technique is a form of regression analysis introduced by Jerome Friedman in 1991 [22]. MARS technique is a non-parametric regression technique and can be seen as an extension of linear models that automatically models non-linearities and interactions. MARS technique [29-31] has been applied greatly in recent years to many fields of science and engineering with success. For instance, MARS technique is used in a variety of fields, including biomedicine and bioinformatics and other engineering fields [21-31]. MARS models are more flexible than linear regression models and they are simple to understand and interpret. MARS technique can handle both numeric and categorical data and it tends to be better than recursive partitioning for numeric data because hinges are more appropriate for numeric variables than the piecewise constant segmentation used by recursive partitioning [20,21]. To fix ideas, building MARS models often requires little or no data preparation. The hinge functions automatically partition the input data, so the effect of outliers is contained. In this respect MARS technique is similar to recursive partitioning which also partitions the data into disjoint regions, although using a different method. MARS models tend to have a good bias-variance trade-off and the MARS models are flexible enough to model non-linearity and variable interactions [28-31].

In this paper, the MARS technique [22-31] was used as an automated learning tool when building three MARS models for nitrogen dioxide (NO_2), sulfur dioxide (SO_2) and aerosol particles less than 10 micrometers (PM_{10}) as a function of other measured relevant pollutants in air quality: nitric oxide (NO), carbon monoxide (CO) and ozone (O_3). The aim was to make accurate concentration estimates of the three above mentioned pollutants (NO_2, SO_2 and PM_{10}) [11,32-34]. MARS models were used as an alternative to the traditional regression approaches. The three MARS models were found to be better to tackle nonlinear regression problems such as those associated to air quality and studied in this research work.

This study is structured as follows: firstly, the materials and methods used to carry out this study are described; next, the obtained results are presented and discussed; and finally, the main conclusions drawn from the results are described.

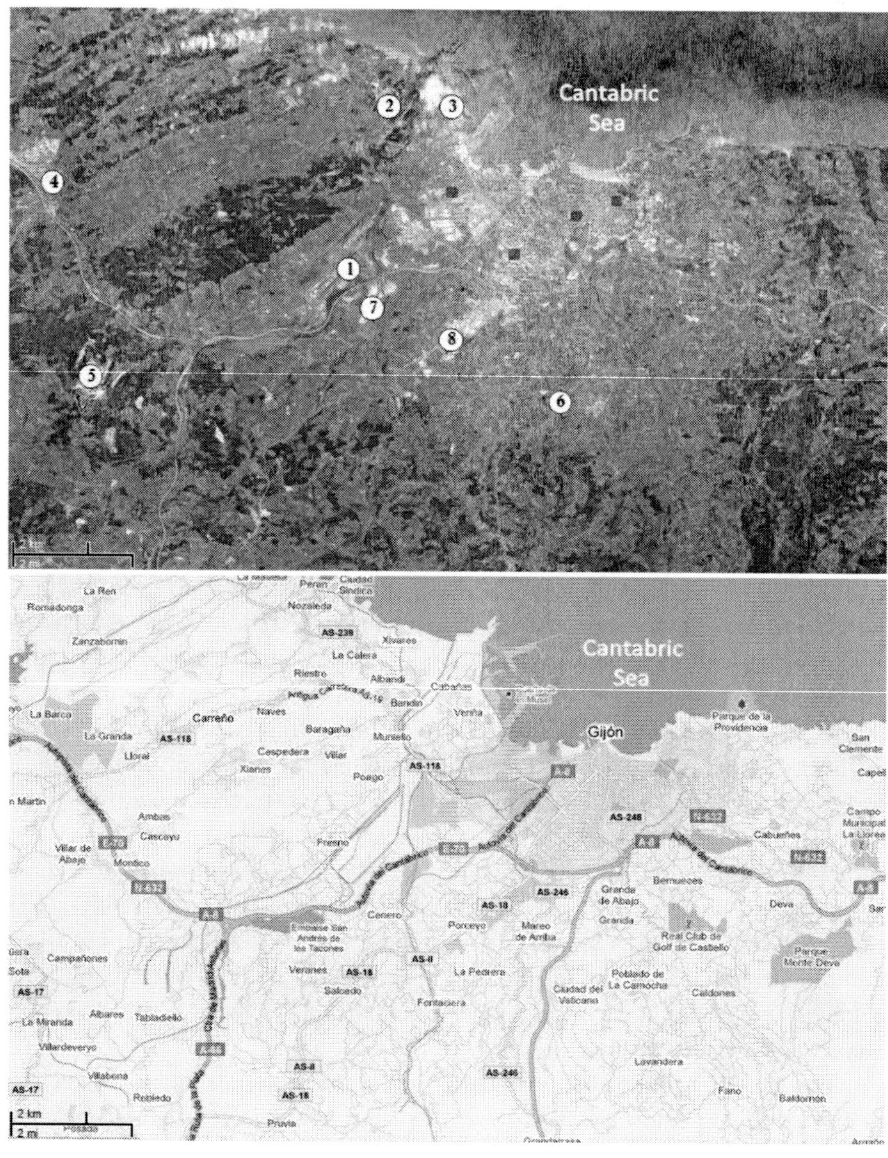

Figure 1. The geographical location (upper) of the air quality stations (red squares) in the city of Gijón (Northern Spain) and the main industries (1: Arcelor Mittal Heavy Steel Industry; 2: Aboño Coal-fired Power Plant; 3: Gijón'sHarbour: El Musel; 4: DuPont Industry; 5: Limestone Quarry; 6: La Camocha Coal Mine; 7: Somonte Industrial Park; 8: Roces Industrial Park), and street map (lower).

2. Materials and Methods

2.1. Sources and Types of Air Pollution

A substance in the air that can be harmful to humans and the environment is known as an air pollutant. Pollutants can be in the form of solid particles, liquid droplets, or gases. In addition, they may be natural or man-made. Pollutants can be classified as primary or secondary. Usually, primary pollutants are directly emitted from a process, such as ash from a volcanic eruption, the carbon monoxide gas from a motor vehicle exhaust or sulphur dioxide released from factories. Secondary pollutants are not emitted directly. Rather, they form in the air when primary pollutants react or interact. An important example of a secondary pollutant is ground level ozone, one of the many secondary pollutants that make up photochemical smog [4,34,35]. Some pollutants may be both primary and secondary: that is, they are both emitted directly and formed from other primary pollutants.

Major primary pollutants produced by human activity include [1-5,11,12,32,33,36,37]:

- Particulatematter (PM): alternatively referred to as atmospheric particulate matter, or fine particles, are tiny particles of solid or liquid suspended in a gas. In contrast, aerosol refers to particles and the gas together. Sources of particulates can be man made or natural. Some particulates occur naturally, originating from volcanoes, dust storms, forest and grassland fires, living vegetation, and sea spray. Human activities, such as the burning of fossil fuels in vehicles, power plants and various industrial processes also generate significant amounts of aerosols. Averaged over the globe, anthropogenic aerosols, those made by human activities, currently account for about 10 percent of the total amount of aerosols in our atmosphere. Increased levels of fine particles in the air are linked to health hazards such as heart disease, altered lung function and lung cancer.
- Sulfur oxides (SO_x): especially sulphur dioxide, a chemical compound with the formula SO_2. SO_2 is produced by volcanoes and in various industrial processes. Since coal and petroleum often contain sulphur compounds, their combustion generates sulfur dioxide. Further oxidation of SO_2, usually in the presence of a catalyst such as NO_2, forms H_2SO_4, and thus acid rain [3-5]. This is one of the causes for

concern over the environmental impact of the use of these fuels as power sources.

- Nitrogen oxides (NO_x): especially nitrogen dioxide are emitted from high temperature combustion, and are also produced naturally during thunderstorms by electric discharge. Can be seen as the brown haze dome above or plume downwind of cities. Nitrogen dioxide is the chemical compound with the formula NO_2. It is one of the several nitrogen oxides. This reddish-brown toxic gas has a characteristic sharp, biting odor. NO_2 is one of the most prominent air pollutants. The initial product formed is nitric oxide (NO). When NO oxidizes further in the atmosphere, nitrogen dioxide (NO_2) forms. Commonly, the general term NO_x is used to describe these gases.
- Carbon monoxide (CO): is a colourless, odorless, non-irritating but very poisonous gas. It is a product by incomplete combustion of fuel such as natural gas, coal or wood. Vehicular exhaust is a major source of carbon monoxide.
- Volatile organic compounds (VOCs): are an important outdoor air pollutant. In this field they are often divided into the separate categories of methane (CH_4) and non-methane (NMVOCs). This pollutant is not considered in this study.

Secondary pollutants considered here include [3-5,9,12,37-40]:

- Particulate matter formed from gaseous primary pollutants and compounds in photochemical smog. Smog is a kind of air pollution. The word smog is a portmanteau of smoke and fog. Classic smog results from large amounts of coal burning in an area caused by a mixture of smoke and sulfur dioxide. Modern smog does not usually come from coal but from vehicular and industrial emissions that are acted on in the atmosphere by ultraviolet light from the sun to form secondary pollutants that also combine with the primary emissions to form photochemical smog.
- Ground level ozone (O_3): formed from NO_x and VOCs. Ozone (O_3) is a key constituent of the troposphere (it is also an important constituent of certain regions of the stratosphere commonly known as the Ozone layer). Photochemical and chemical reactions involving it drive many of the chemical processes that occur in the atmosphere by day and by night. At abnormally high concentrations brought about by human activities (largely the combustion of fossil fuel), it is a pollutant, and a

constituent of smog. The negative effects of ozone are well documented. Short-term exposure to elevated levels of ozone causes eye and lung irritations [40].

With respect to the trends of in air quality, the Clean Air Act of 1970 mandated the setting of standards for four of the primary pollutants (aerosols, sulphur dioxide, carbon monoxide, and nitrogen oxides) as well as the secondary pollutant ozone. At the time, these five pollutants were recognized as being the most widespread and objectionable. Today, with the addition of lead, they are known as the criteria pollutants and are covered by the United States National Ambient Air Quality Standards (see Table 1 below) [3,9,11]. The primary standard for each pollutant shown in Table 1 is based on the highest level that can be tolerated by humans without noticeable ill effects, minus a 10-50% margin for safety reasons.

Table 1. National ambient air quality standards by United States Environmental Protection Agency (USEPA) [3,9,11]

Pollutant	Maximum allowable concentrations
Carbon monoxide (CO)	
8-hour average	9 ppm (10 mg/m^3)
1-hour average	35 ppm (40 mg/m^3)
Nitrogen dioxide (NO_2)	
Annual arithmetic mean	0.053 ppm (100μg/m^3)
Ozone (O_3)	
1-hour average	0.12 ppm (235μg/m^3)
8-hour average	0.08 ppm (157μg/m^3)
Particulate <10 micrometers (PM_{10})	
Annual arithmetic mean	50μg/m^3
24-hour average	150μg/m^3
Particulate <2.5 micrometers ($PM_{2.5}$)	
Annual arithmetic mean	15μg/m^3
24-hour average	65μg/m^3
Sulphur dioxide (SO_2)	
Annual arithmetic mean	0.03 ppm (80μg/m^3)
24-hour average	0.14 ppm (365μg/m^3)

2.2. Experimental Dataset

The Section of Industry and Energy from the government of Asturias has four meteorological stations distributed throughout the city of Gijón (see Figure 1 above). These four stations measure every fifteen minutes the following primary and secondary pollutants: sulphur dioxide (SO_2), nitrogen oxides (NO and NO_2), carbon monoxide (CO), particulate matter less than 10 micrometers (PM_{10}) and ozone (O_3). This data set is collected, processed and delivered on average for the entire city every day. Therefore, we have data for the pollutants listed above each day from January 2006 to December 2008.

Then, it is possible to study the trend in emissions of the preceding pollutants in the years 2006, 2007 and 2008 [4,5,9,34]. In the first place, Figure 2 (a) shows the sulphur dioxide emissions each month during the years 2006, 2007 and 2008. Emissions of this pollutant are kept in a band or range of concentrations between approximately 7 and 27 mg/m^3, regardless of the month and the season, reaching the maximum emission during the Christmas of 2007 (26.53 mg/m^3) and the minimum emission (7.16 mg/m^3) in July 2007. This trend is general throughout the years studied, and it is within the logic, as one coal-fired power plant is close to this area and another important source of this gas is the heavy steel industry, in which the production rate is more or less stable throughout the year (see Figure 1 above). From the standard air quality point of view, following the USEPA Air Quality Standards (see Table 1 above), the maximum allowable concentration of SO_2 expressed as annual arithmetic mean is 80 mg/m^3. The annual arithmetic means for this gas during the years 2006, 2007 and 2008 were 12.85, 14.23 and 10.57 mg/m^3 respectively. Therefore, the emissions of this gas were much lower than the maximum permitted and meet air quality standards for a healthy person during these three years, including emission peaks.

Secondly, Figure 2 (a) shows the nitric oxide emissions each month during the years 2006, 2007 and 2008. It is possible to observe that the emission peaks occur during late autumn and early winter, namely from October to February each year, reaching the maximum emission in December 2007: 68.16 μg/m^3. Similarly minimum emissions (12.05 μg/m^3 in April 2006) occur more often during the summer months. This trend is a quasi-sinusoidal curve with peaks and valleys. Although the initial product of the combustion is nitric oxide (NO), this gas is rapidly oxidized and converted to NO_2. Its residence time in the atmosphere is very short and the USEPA Air Quality Standards does not consider it.

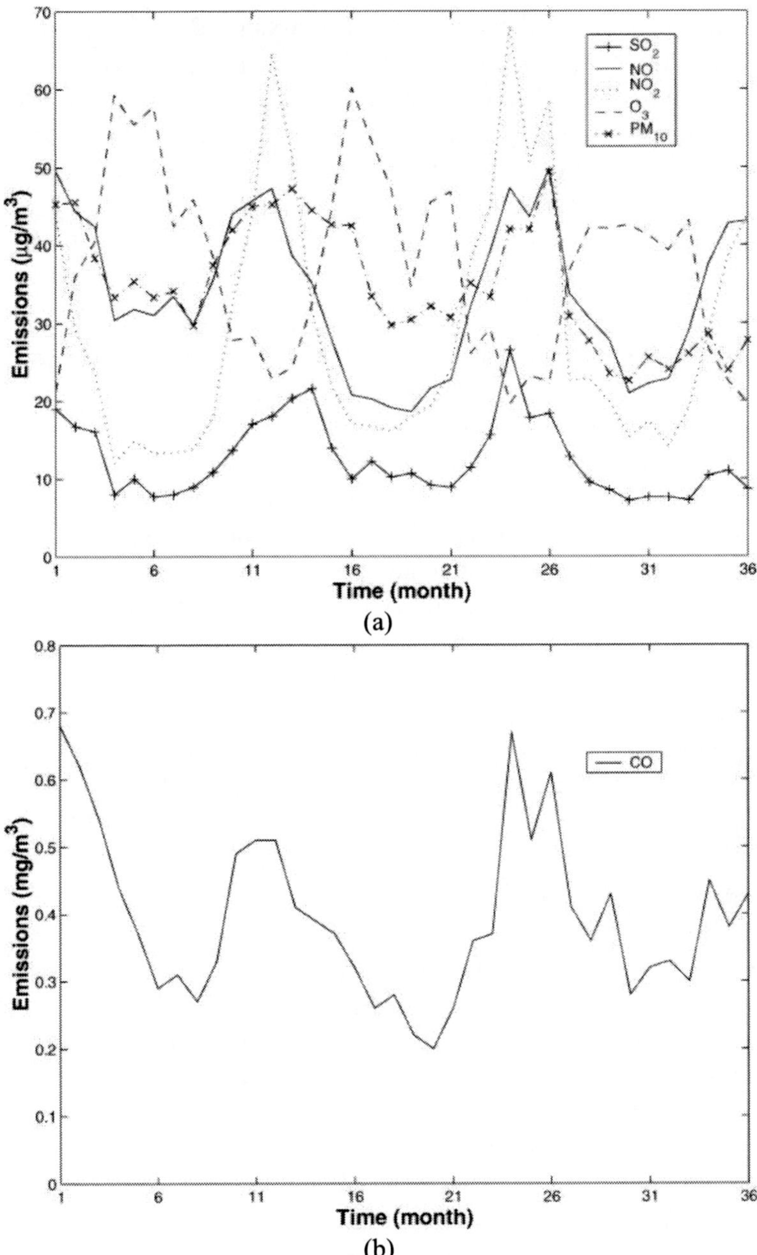

Figure 2. (a) Monthly trend of sulphur dioxide (SO_2), nitric oxide (NO), nitrogen dioxide (NO_2), ozone (O_3) particulate matter (PM_{10}); and (b) carbon monoxide (CO) emissions (right) during the years 2006, 2007 and 2008 inGijón urban area.

Thirdly, the nitrogen dioxide (NO_2) emissions each month during the years 2006, 2007 and 2008 are also shown in Figure 2 (a). The trend of this gas is similar to that observed for the previous gas throughout the years studied, although more pronounced. Note that the emission peaks occur from September to December or January each year, reaching the maximum emission in February 2008: 49.95 $\mu g/m^3$. The minimum emission occurred in July 2007 (18.61 $\mu g/m^3$) and this trend is maintained, as the minimum values for the rest of the years occurred in August 2006 (29.81 $\mu g/m^3$) and June 2008 (20.96 $\mu g/m^3$). Following the USEPA Air Quality Standards (see Table 1 above), the maximum allowable concentration of NO_2 expressed as annual arithmetic mean is 100 $\mu g/m^3$. The annual arithmetic means for this gas during the years 2006, 2007 and 2008 were 38.81, 28.63 and 33.69 mg/m^3 respectively. Thus, the nitrogen dioxide (NO_2) emissions are also below the maximum permitted and meet air quality standards for a healthy person during these three years, including emission peaks. As mentioned previously, it is also important to highlight a certain flattening of the emissions of this gas in time.

Fourthly, Figure 2 (b) also shows the carbon monoxide (CO) emissions each month during the years 2006, 2007 and 2008. The main emission peaks occur in January 2006 and December 2008 with a value of 0.68 and 0.67 $\mu g/m^3$, respectively. Minimum emissions occur in June (0.29 $\mu g/m^3$) and August 2006 (0.27 $\mu g/m^3$), in July (0.22 $\mu g/m^3$) and August 2007 (0.20 $\mu g/m^3$), and in June (0.28 $\mu g/m^3$) and September 2008 (0.30 $\mu g/m^3$). The quasi-sinusoidal trend is lost here. Similarly, following the USEPA Air Quality Standards [1-3,9] (see Table 1 above), the maximum allowable concentration of CO expressed as annual arithmetic mean is 3.33 $\mu g/m^3$. The annual arithmetic means for this gas during the years 2006, 2007 and 2008 were 0.45, 0.34 and 0.40 $\mu g/m^3$ respectively. Hence the emissions of CO were below the highest level that can be tolerated by humans according to USEPA Air Quality Standards during these three years, including emission peaks.

Fifthly, Figure 2 (a) also shows the particulate matter PM_{10} emissions each month during the years 2006, 2007 and 2008. It is possible to observe that the emission peaks occur in January and February 2006, January 2007 and February 2008 with values 45.31 $\mu g/m^3$, 45.59 $\mu g/m^3$, 47.26 $\mu g/m^3$ and 49.36 $\mu g/m^3$, respectively. As for the previous pollutant, the quasi-sinusoidal trend is less lost here in the case of particulate matter. All aerosol emissions are kept in a band or range of concentrations between approximately 22 and 50 mg/m^3, regardless of the month and the season. From the point of standard air quality view, following the USEPA Air Quality Standards (see Table 1 above), the maximum allowable concentration of PM_{10} expressed as annual arithmetic mean is 50 μg/m^3. The annual arithmetic means for this pollutant during the years 2006, 2007 and 2008 were 38.75, 37.00 and 29.33 μg/m^3 respectively. Therefore, the aerosol emissions are below the allowable maximum for a healthy person during these three years, although emission peaks are close to this value. This behaviour can give place to serious health problems for the population such as chronic diseases and deaths.

Figure 2 (a) also shows the ozone emissions each month during the years 2006, 2007 and 2008. Note that there is a variation of the ozone concentration in the form of an oscillating sawtooth in time, reaching maximum values during spring and summer months: 59.28 μg/m^3 in April 2006, 57.83 mg/m^3 in June 2006, 60.21 μg/m^3 in April 2007, 53.41 μg/m^3 in May 2007, 47.27 μg/m^3 in June 2007, 46.83 μg/m^3 in September 2007, 42.57μg/m^3 in June 2008 and 43.22μg/m^3 in September 2008. This trend is general throughout the years studied, since ozone is associated with photochemical reactions, and these ones require the presence of strong sunlight. Moreover, the evolution of the peaks and valleys in this gas is opposite that of NO and NO_2, especially the latter, which is consistent with the photochemical behaviour and the relationship between these two pollutants. The Clean Air Act directs the USEPA to set National Ambient Air Quality Standards for several pollutants, including ground-level ozone, and cities out of compliance with these standards are required to take steps to reduce their levels. In May 2008, the USEPA lowered its ozone standard from 80 μg/m^3 to 75 μg/m^3. This proved controversial, since the Agency's own scientists and advisory board had recommended lowering the standard to 60 μg/m^3, and the World Health Organization recommends 51 μg/m^3. Many public health and environmental groups also supported the 60 μg/m^3 standard. The annual arithmetic means for this gas in Gijón urban nucleus during the years 2006, 2007 and 2008 were 39.69, 38.69 and 33.48 μg/m^3, respectively. Therefore, the emissions of this gas were below the maximum permitted and meet air quality standards during these three years, including emission peaks. However, in April 2007, a peak was reached above 60 μg/m^3 recommendation. This fact could be dangerous for the health of the population of Gijón. There is a great deal of evidence to show that high concentrations of ozone, created by high concentrations of pollution and daylight UV rays at the Earth's surface, can harm lung function and irritate the respiratory system. Exposure to ozone and the pollutants that produce it has been linked to premature death, asthma, bronchitis, heart attack, and other cardiopulmonary problems. According to scientists of the United

States Environmental Protection Agency (USEPA), susceptible people can be adversely affected by ozone levels as low as 40 μg/m^3.

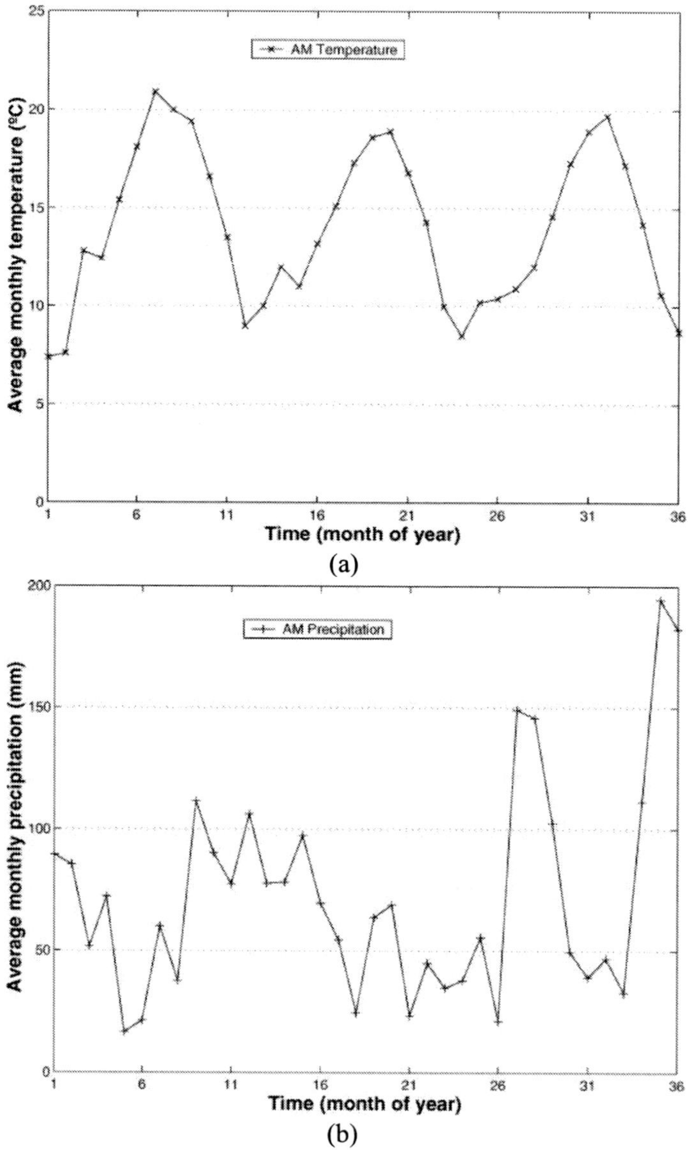

Figure 3. Meteorological data: (a) average monthly temperature and (b) average monthly precipitation in Gijón urban nucleus from January 2006 (month 1) to December 2008 (month 36).

Finally, in order to explain the peaks of NO_2, NO and CO emissions during the period studied, it is necessary to take into account the role of meteorology. In this sense, Figure 3 (a) shows the average monthly temperature and Figure 3 (b) the average monthly precipitation in Gijón urban nucleus from January 2006 to December 2008. It is possible to observe a relation between the coldest and wettest months and the peaks of the pollutants due to the more intense use of vehicles and heating. Similarly, in the case of ozone an inverse relationship can be observed between the level of emissions and rainfall.

2.3. Multivariate Adaptive Regression Splines Method (MARS)

Multivariate adaptive regression splines (MARS) is a multivariate nonparametric classification/regression technique introduced by Friedman [22,24]. The theoretical model that is explained below has already been presented by the authors in previous researches [30,31]. In spite of this fact and due to its interest for the reader in order to achieve a full understanding of the research that is presented in this paper. Its main purpose is to predict the values of a continuous dependent variable, **y** ($n \times 1$), from a set of independent explanatory variables, **X** ($n \times p$). The MARS model can be represented as:

$$\mathbf{y} = f(\mathbf{X}) + \mathbf{e} \tag{1}$$

where f is a weighted sum of basis functions that depend on **X** and **e** is an error vector of dimension ($n \times 1$).

MARS model does not require any a priori assumptions about the underlying functional relationship between dependent and independent variables. Instead, this relation is uncovered from a set of coefficients and piecewise polynomials of degree q (basis functions) that are entirely “driven” from the regression data (**X**, **y**). The MARS regression model is constructed by fitting basis functions to distinct intervals of the independent variables. Generally, piecewise polynomials, also called splines, have pieces smoothly connected together. In MARS terminology, the joining points of the polynomials are called knots, nodes or breakdown points. These will be denoted by the small letter t. For a spline of degree q each segment is a polynomial function. MARS uses two-sided truncated power functions as spline basis functions, described by the following equations [22-31]:

$$[-(x-t)]_+^q = \begin{cases} (t-x)^q & \text{if } x < t \\ 0 & \text{otherwise} \end{cases} \tag{2}$$

$$[+(x-t)]_+^q = \begin{cases} (t-x)^q & \text{if } x \geq t \\ 0 & \text{otherwise} \end{cases} \tag{3}$$

where $q(\geq 0)$ is the power to which the splines are raised and which determines the degree of smoothness of the resultant function estimate. When $q = 1$, which is the case in this study, only simple linear splines are considered.

The MARS model of a dependent variable $\mathbf{y}$ with M basis functions (terms) can be written as [22,24,28-31]:

$$\hat{\mathbf{y}} = \hat{f}_M(\mathbf{x}) = c_0 + \sum_{m=1}^{M} c_m B_m(\mathbf{x}) \tag{4}$$

where $\hat{\mathbf{y}}$ is the dependent variable predicted by the MARS model, c_0 is a constant, $B_m(\mathbf{x})$ is the m-th basis function, which may be a single spline basis functions, and c_m is the coefficient of the m-th basis functions.

Both the variables to be introduced into the model and the knot positions for each individual variable have to be optimized. For a data set $\mathbf{X}$ containing n objects and p explanatory variables, there are $N = n \times p$ pairs of spline basis functions, given by Eqs. (2) and (3), with knot locations x_{ij} ($i = 1,2,...,n$; $j = 1,2,..., p$).

A two-step procedure is followed to construct the final model. First, in order to select the consecutive pairs of basis functions of the model, a two-at-a-time forward stepwise procedure is implemented [22,24-31]. This forward stepwise selection of basis function leads to a very complex and overfitted model. Such a model, although it fits the data well, has poor predictive abilities for new objects. To improve the prediction, the redundant basis functions are removed one at a time using a backward stepwise procedure. To determine which basis functions should be included in the model, MARS utilizes the generalized cross-validation (*GCV*) [22,28-31]. In this way, the *GCV* is the mean squared residual error divided by a penalty dependent on the model complexity. The *GCV* criterion is defined in the following way [22,24,30,31]:

$$GCV(M) = \frac{\frac{1}{n}\sum_{i=1}^{n}\left(y_i - \hat{f}_M(\mathbf{x}_i)\right)^2}{\left(1 - C(M)/n\right)^2} \tag{5}$$

where $C(M)$ is a complexity penalty that increases with the number of basis functions in the model and which is defined as [22-31]:

$$C(M) = (M+1) + dM \tag{6}$$

where M is the number of basis functions in Eq. (4), and the parameter d is a penalty for each basis function included into the model. It can be also regarded as a smoothing parameter. Large values of d lead to fewer basis functions and therefore smoother function estimates. For more details about the selection of the d parameter, see the references [22,24,28-31]. In our studies, the parameter d equals 2, and the maximum interaction level of the spline basis functions is restricted to 3.

2.3.1. The importance of the variables in the MARS model

Once the MARS model is constructed, it is possible to evaluate the importance of the explanatory variables used to construct the basis functions. Establishing predictor importance is in general a complex problem which in general requires the use of more than one criterion. In order to obtain reliable results, it is convenient the use of the *GCV* parameter explained before together with the parameters Nsubsets (criterion counts the number of model subsets in which each variable is included) and the residual sum of squares *RSS* [24-31].

3. ANALYSIS OF RESULTSAND DISCUSSION

The list of input variables taken into account in this research work is shown in Table 2. The total number of dependent variables used to build the MARS models was 3: nitrogen dioxide (NO_2), sulfur dioxide (SO_2) and particulate matter less than 10μm (PM_{10}). Indeed, we have built three different MARS models taking as dependent variables NO_2, SO_2 and PM_{10}, respectively; and as independent input variables (predictor variables) the other variables listed in Table 2.

In this work, three second-order MARS models have been used, so that the basis functions of the model consist of linear and second-order splines and the maximum number of terms was not limited (no pruning). The results of the MARS models computed using all the available data observations are shown in Tables 3, 4 and 5. Tables 3, 4 and 5 show a list of 4, 6 and 4 main basis functions for each of the three MARS models and their coefficients, respectively. Please note that $h(x) = x$ if $x > 0$ and $h(x) = 0$ if $x \leq 0$. Therefore, the MARS model is a form of non-parametric regression technique and can be seen as an extension of linear models that automatically models non-linearities and interactions as a weighted sum of basis functions called *hinge functions* [22,24-31]. The predicted response for NO_2, SO_2 and PM_{10} is now a better fit to the original values since the MARS model has automatically produced a kink in the predicted dependent variable to take into account non-linearities.

According to the results shown in Table 6, the most important variables for the prediction of the NO_2 (output variable) are as follows:ozone (O_3) and finally carbon monoxide (CO). The remaining input variables (SO_2, NO and PM_{10}) were negligible according to the calculations carried out through the first MARS model.

Table 2. Set of input variables used in this study

Input variables	Name of the variable
SO_2(μg/m^3)	Sulfur dioxide
NO (μg/m^3)	Nitric oxide
NO_2(μg/m^3)	Nitrogen dioxide
CO (mg/m^3)	Carbon monoxide
PM_{10}(μg/m^3)	Particles less than 10 μm
O_3(μg/m^3)	Ozone

Table 3. List of basis functions of the MARS model for the NO_2 pollutant and their coefficients c_i

B_i	Definition	c_i
B_1	1	31.7173
B_2	h(CO – 0.39)	38.4829
B_3	h(0.39 – CO)	– 67.7779
B_4	h(41.19 – O_3)	0.436871

Table 4. List of basis functions of the MARS model for the SO_2 pollutant and their coefficients c_i

B_i	Definition	c_i
B_1	1	8.9236
B_2	$h(NO - 21.8)$	0.1657
B_3	$h(CO - 0.43)$	39.7790
B_i	Definition	c_i
B_4	$h(PM_{10} - 35.11)$	0.9629
B_5	$h(NO_2 - 35.19) \times h(PM_{10} - 35.11)$	– 0.0884
B_6	$h(35.19 - NO_2) \times h(PM_{10} - 35.11)$	– 0.0533

Table 5. List of basis functions of the MARS model for the PM_{10} pollutant and their coefficients c_i

B_i	Definition	c_i
B_1	1	42.2358
B_2	$h(13.94 - SO_2)$	– 3.2426
B_3	$h(0.39 - CO)$	0.0902
B_4	$h(13.94 - SO_2) \times h(O_3 - 36.69)$	0.4369

Table 6. Evaluation of the importance of the variables that form the model for the NO_2 pollutant according to criteria Nsubsets, *GCV* and *RSS*

Input variables	Nsubsets	*GCV*	*RSS*
O_3	3	100.0	100.0
CO	2	51.0	52.3

Similarly, from the results shown in Table 7, it is possible to observe that the most important variables for the prediction of the SO_2 (output variable) are in hierarchical order:PM_{10}, NO, NO_2 and CO. The remaining input variable (O_3) was negligible according to the calculations carried out through this second MARS model.

Table 7. Evaluation of the importance of the variables that form the model for the SO_2 pollutant according to criteria Nsubsets, *GCV* and *RSS*

Input variables	Nsubsets	*GCV*	*RSS*
PM_{10}	5	100.0	100.0
NO	4	48.6	54.9
NO_2	2	36.2	35.7
CO	2	36.2	35.7

Additionally, the results shown in Table 8 indicate that the most important variables for the prediction of the PM_{10} (output variable) are mainly:SO_2 and finally O_3. The remaining input variables (NO, NO_2 and CO) were negligible according to the calculations carried out through this third MARS model.

Table 8. Evaluation of the importance of the variables that form the model the PM_{10} pollutant according to criteria Nsubsets, *GCV* and *RSS*

Input variables	Nsubsets	*GCV*	*RSS*
SO_2	2	100	100
O_3	1	32.3	37.5

Furthermore, a graphical representation of the terms that constitute the three MARS models can be seen in Figures 4, 5 and 6, respectively.

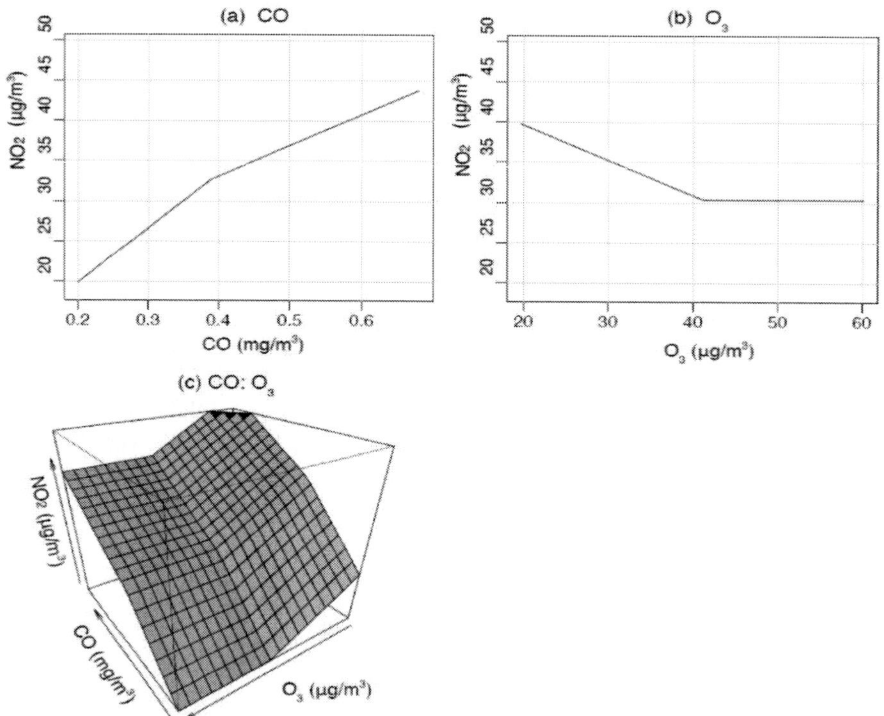

Figure 4. Graphical representation of the terms that constitute the MARS model for the NO_2 pollutant: (a) first order term of the predictor variable CO; (b) first order term of the predictor variable O_3; and (c) second order term of the predictor variables CO and O_3 values.

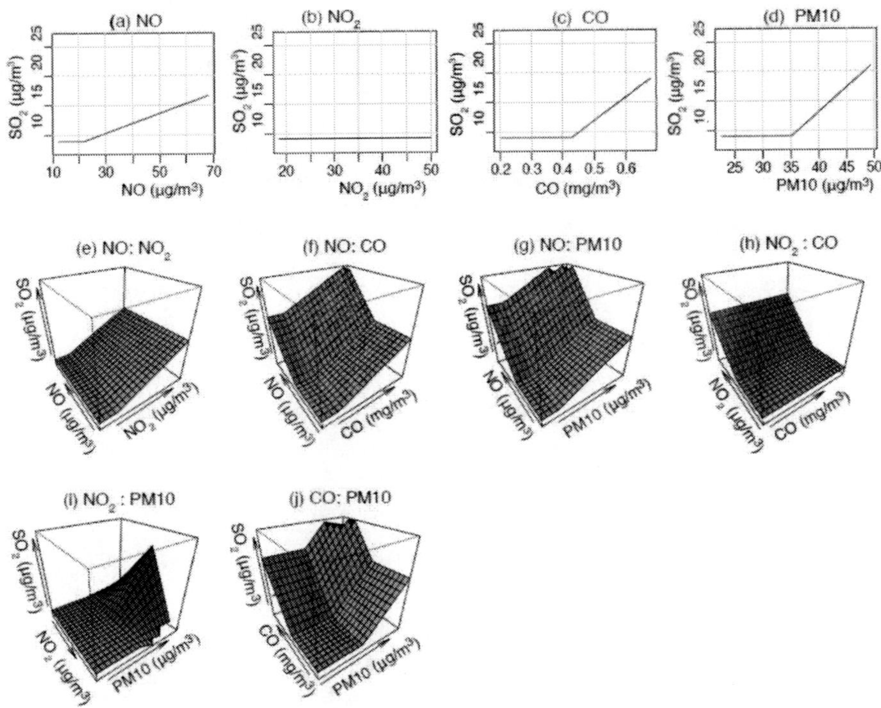

Figure 5. Graphical representation of the terms that constitute the MARS model for the SO_2 pollutant: (a) first order term of the predictor variable NO; (b) first order term of the predictor variable NO_2; (c) first order term of the predictor variable CO; (d) first order term of the predictor variable PM_{10}; (e) second order term of the predictor variables NO and NO_2 values; (f) second order term of the predictor variables NO and CO values; (g) second order term of the predictor variables NO and PM_{10} values; (h) second order term of the predictor variables NO_2 and CO values; (i) second order term of the predictor variables NO_2 and PM_{10} values; and (j) second order term of the predictor variables CO and PM_{10} values.

It is very important to select the model that best fits the experimental data. The goodness of fit of a statistical model describes how well it fits a set of observations. Measures of goodness of fit typically summarize the discrepancy between observed values and the values expected under the model in question. In regression analysis, the following criterion was considered in this research relates to goodness of fit: the coefficient of determination R^2 [28,30,31,41]. This ratio indicates the proportion of total variation in the dependent variables (NO_2, SO_2 and PM_{10} in our case) explained by the MARS models. A dataset takes values t_i, each of which has an associated modelled value y_i. The former are called the observed values and the latter are often referred to as the predicted values. Variability in the dataset is measured through different sums of squares:

- $SS_{tot} = \sum_{i=1}^{n} \left(t_i - \bar{t}\right)^2$: the total sum of squares, proportional to the sample variance.
- $SS_{reg} = \sum_{i=1}^{n} \left(y_i - \bar{t}\right)^2$: the regression sum of squares, also called the explained sum of squares.
- $SS_{err} = \sum_{i=1}^{n} \left(t_i - y_i\right)^2$: the residual sum of squares.

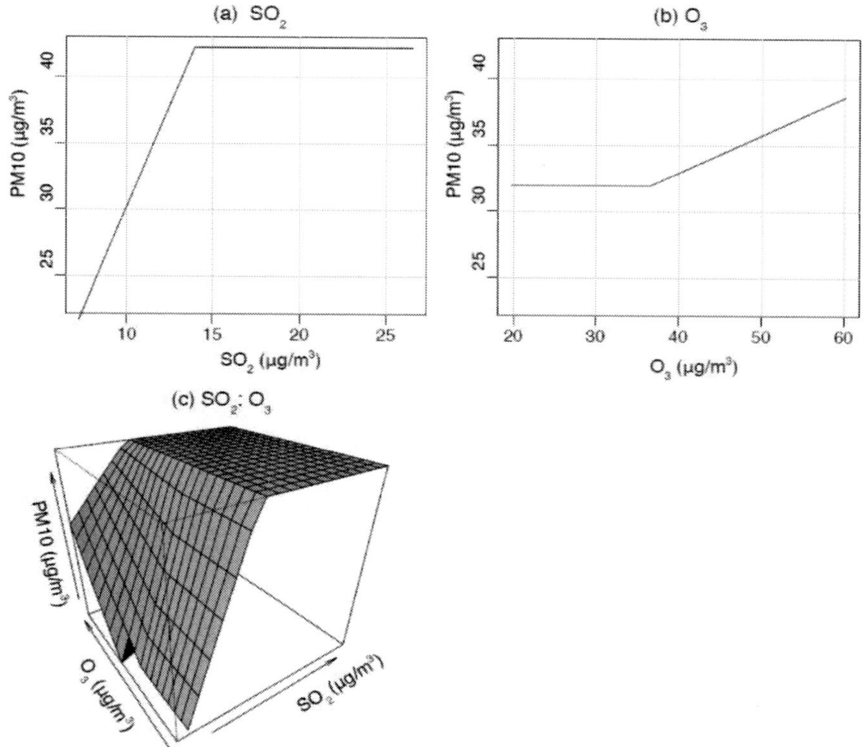

Figure 6. Graphical representation of the terms that constitute the MARS model for the PM_{10} pollutant: (a) first order term of the predictor variable SO_2; (b) first order term of the predictor variable O_3; and (c) second order term of the predictor variables SO_2 and O_3values.

In the previous sums, $\bar{t}$ is the mean of the n observed data:

$$\bar{t} = \frac{1}{n} \sum_{i=1}^{n} t_i \tag{7}$$

Bearing in mind the above sums, the general definition of the coefficient of determination is:

$$R^2 \equiv 1 - \frac{SS_{err}}{SS_{tot}} \tag{8}$$

A coefficient of determination value of 1.0 indicates that the regression curve fits the data perfectly. In this current research work, the three fitted MARS models for NO_2, SO_2 and PM_{10} have coefficients of determination equal to 0.84, 0.90 and 0.77, respectively. Addtionally, their correlation coefficients were 0.92, 0.95 and 0.88, respectively. These results indicate a very high goodness of fit for three MARS models analysed.

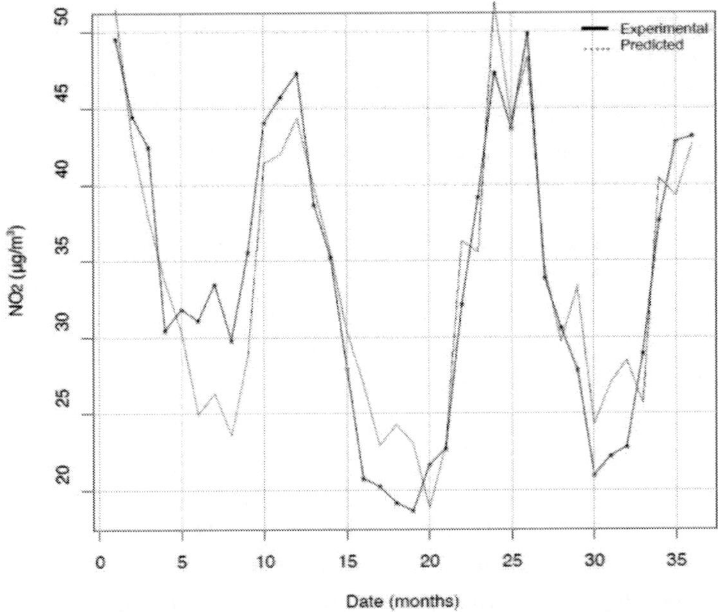

Figure 7. Comparison between the concentrations of NO_2 observed and predicted by the model MARS from 2006 to 2008.

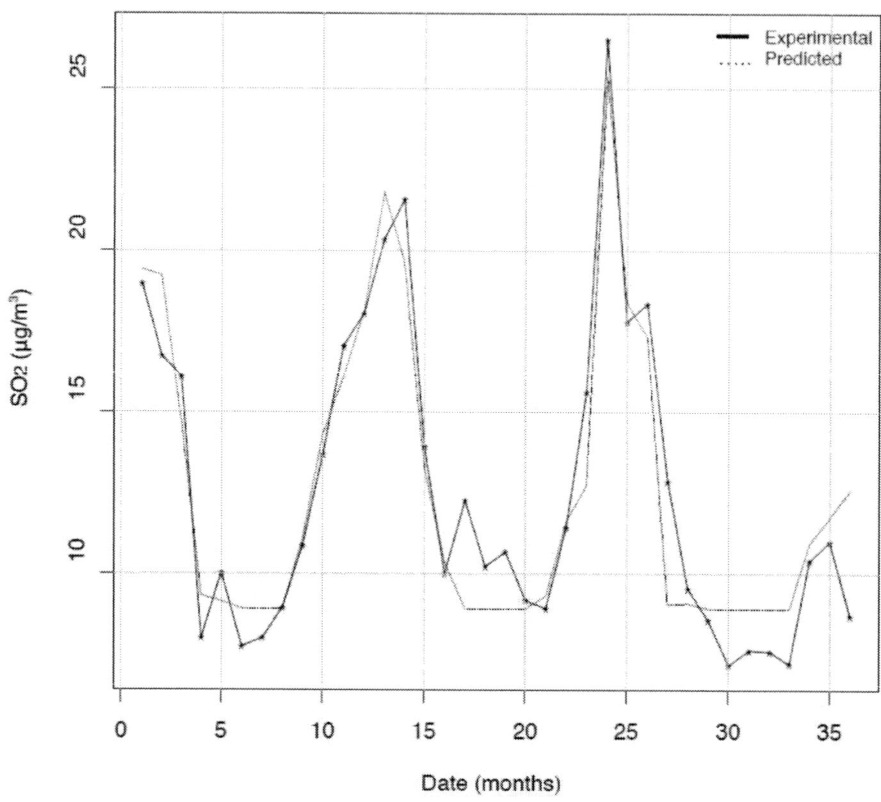

Figure 8. Comparison between the concentrations of SO_2 observed and predicted by the model MARS from 2006 to 2008.

In order to guarantee the ability prediction of the three MARS models, the cross validation [42,43] was the standard technique used here for finding a suitable set of hyperparameters of the three MARS models built in this research work. In this sense, the data set is randomly divided into l disjoint subsets of equal size, and each subset is used once as a validation set, whereas the other $l - 1$ subsets are put together to form a training set. In the simplest case, the average accuracy of the l validation sets is used as an estimator for the accuracy of the method. In this research work, 10-fold cross-validation was used, that is to say, to calculate the error criterion, the models were built using 90% of the sample and tested with the remaining 10%, thus simulating as closely as possibly the real conditions under which the model would be built in order to later fit it to new observation data unrelated to the construction of the models.

Finally, this research work was able to estimate the concentrations of NO_2 from 2006 to 2008 in agreement to the experimental actual concentrations of NO_2 observed with success (see Figure 7). Similarly, Figures 8 and 9 show a good agreement between the experimental concentrations of SO_2 and PM_{10} and their predicted concentrations using the MARS models from 2006 to 2008, respectively.

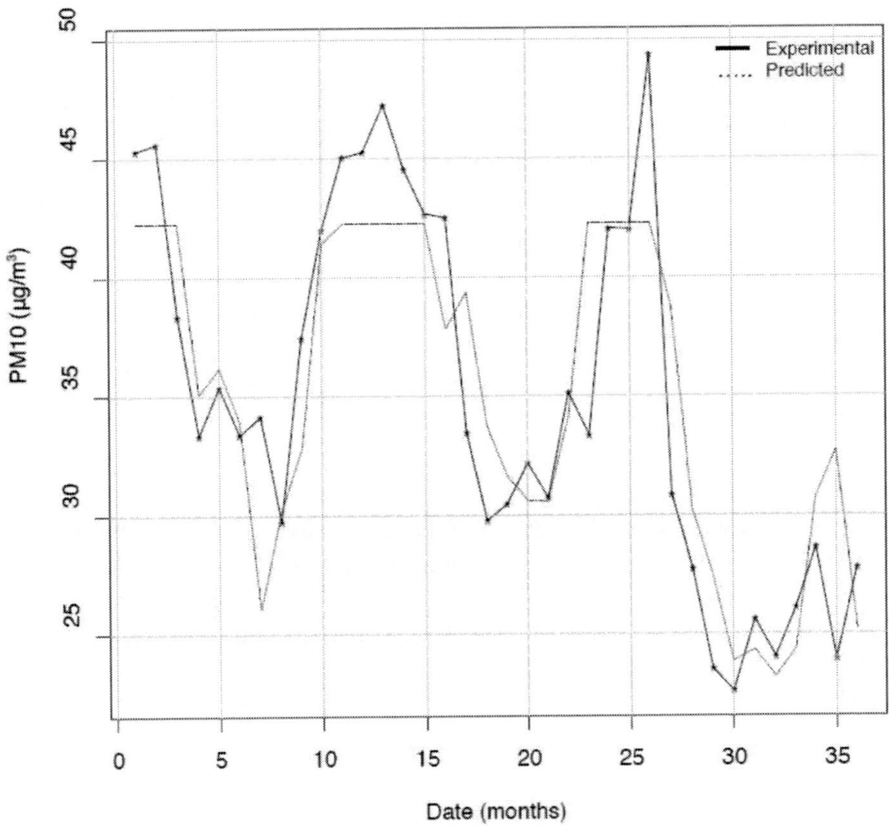

Figure 9. Comparison between the concentrations of PM_{10} observed and predicted by the model MARS from 2006 to 2008.

Conclusion

In the first place, this research described steps for the construction of three MARS models to estimate quickly and with a high degree of accuracy the concentrations of NO_2, SO_2 and PM_{10} from 2006 to 2008. We have provided examples of real applications and simple explanations of two commonly used statistics for the selection of the best-fitting models: the coefficient of determination and correlation coefficient.

Secondly, the MARS model is potentially useful for predicting pollutant concentrations in the atmosphere. In other words, this new and innovative methodology developed here could be applied to other industrial cities with similar or different sources of pollutants, but it would be necessary to take into account the specific nature of each location.

Finally, the results of this research about the development of models of local pollutant concentrations is a valuable tool for mitigation projects of acid rain and for the research of effect of particulate matter on human health. Additionally, one of the main findings of this study was to set the order of priority (hierarchy) of the predictor variables involved in the estimation of the dependent variables: NO_2, SO_2 and PM_{10}. Furthermore, this paper presents examples of real applications and simple explanations of statistical calculation for the selection of the best fit models.

Acknowledgments

The authors wish to acknowledge the computational support provided by the Department of Mathematics at University of Oviedo. Additionally, authors would like to express their gratitude to the Department of Education and Science of the Principality of Asturias for its partial financial support (Grant reference FC-11-PC10-19). Finally, we would like to thank Anthony Ashworth for his revision of English grammar and spelling of the manuscript.

References

[1] García Nieto, P.J. (2001). Parametric study of selective removal of atmospheric aerosol by coagulation, condensation and gravitational settling. *International Journal of Environmental Health Research*, *11*, 151–162.

[2] García Nieto, P.J. (2006). Study of the evolution of aerosol emissions from coal-fired power plants due to coagulation, condensation, and gravitational settling and health impact. *Journal of Environmental Management*, *79* (4), 372–382.

[3] Lutgens, F.K. &Tarbuck, E.J. (2001). *The Atmosphere: An Introduction to Meteorology*. New York: Prentice Hall.

[4] Wark, K., Warner, C.F. & Davis, W.T. (1997). *Air Pollution: Its Origin and Control*. New York: Prentice Hall.

[5] Wang, L.K., Pereira, N.C. & Hung, Y.T. (2004). *Air Pollution Control Engineering*. New York: Humana Press.

[6] Karaca, F., Alagha, O. &Ertürk, F. (2005). Statistical characterization of atmospheric PM_{10} and $PM_{2.5}$ concentrations at a non-impacted suburban site of Istanbul, Turkey. *Chemosphere*, *59* (8), 1183–1190.

[7] Comrie, A.C. & Diem, J.E. (1999). Climatology and forecast modeling of ambient carbon monoxide in Phoenix. *Atmospheric Environment,33*, 5023–5036.

[8] Elbir, T., Muezzinoglu, A. &Bayram A. (2000). Evaluation of some air pollution indicators in Turkey. *Environment International*, *26* (1-2), 5–10.

[9] Godish, T. (2004).*Air Quality*. Boca Ratón, Florida: Lewis Publishers.

[10] Akkoyunku, A. &Ertürk, F.A. (2003). Evaluation of air pollution trends in Istanbul. *International Journal of Environmental Pollution*, *18*, 388–398.

[11] Suárez Sánchez, A., García Nieto, P.J., Riesgo Fernández, P., del Coz Díaz, J.J. & Iglesias-Rodríguez, F.J. (2011). Application of a SVM-based regression model to the air quality study at local scale in the Avilés urban area (Spain). *Mathematical and Computer Modelling*, *54* (5–6), 1453–1466.

[12] Cooper, C.D. & Alley, F.C. (2002). *Air Pollution Control*. New York: Waveland Press.

[13] Boznar, M., Lesjack, M. &Mlakar, P. (1993). A neural network based method for short-term predictions of ambient SO2 concentrations in highly polluted industrial areas of complex Terrain. *Atmospheric Environment*, *270*, 221–230.

[14] Haykin, S. (1999). *Neural Networks:Comprehensive Foundation*. New Jersey: Prentice Hall.

[15] Hooyberghs, J., Mensink, C., Dumont, D., Fierens, F. &Brasseur, O. (2005). A neural network forecast for daily average PM_{10} concentrations in Belgium. *Atmospheric Environment*, *39* (18) (2005) 3279–3289.

[16] Kukkonen, J., Partanen, L., Karpinen, A., Ruuskanen, J., Junninen, H., Kolehmainen, M., Niska, H., Dorling, S., Chatterton, T., Foxall, R. &Cawley, G. (2003). Extensive evaluation of neural networks models for the prediction of NO_2 and PM_{10} concentrations, compared with a

deterministic modelling system and measurements in central Helsinki. *Atmospheric Environment*, *37*, 4539–4550.

[17] Gardner, M.W. & Dorling, S.R. (1999). Neural network modelling and prediction of hourly NO_x and NO_2 concentrations in urban air in London. *Atmospheric Environment*, *33* (5), 709–719.

[18] Chaloulakou, A., Saisana, M. &Spyrellis, N. (2003). Comparative assessment of neural networks and regression models for forecasting summertime ozone in Athens. *Science of Total Environment*, *313*, 1–13.

[19] Karaca, F., Nikov, A. &Alagha, O. (2006). NN-AirPol: a neural-network-based method for air pollution evaluation and control. *International Journal of Environmental Pollution*, *28* (3-4), 310–325.

[20] Bishop, C.M. (2006). *Pattern Recognition and Machine Learning*. New York: Springer.

[21] Vapnik, V. (1999). *The Nature of Statistical Learning Theory*. New York: Springer.

[22] Friedman, J.H. (1991). Multivariate adaptive regression splines (with discussion). *Annals of Statistics*, *19*, 1–141.

[23] Hastie, T., Tibshirani, R. & Friedman, J. (2009). *The Elements of Statistical Learning: Data Mining, Inference, and Prediction*. New York: Springer.

[24] Friedman, J.H. &Roosen, C.B. (1995). An introduction to multivariate adaptive regression splines. *Statistical Methods in Medical Research*, *4*, 197–217.

[25] Sekulic, S.S. & Kowalski, B.R. (1992). MARS: A tutorial. *Journal of Chemometrics*, *6*, 199–216.

[26] Xu, Q.S., Daszykowski, M., Walczak, B., Daeyaert, F., de Jonge, M.R., Heeres, J., Koymans, L.M.H., Lewi, P.J., Vinkers, H.M., Janssen, P.A. &Massart, D.L. (2004). Multivariate adaptive regression splines—studies of HIV reverse transcriptase inhibitors. *Chemometrics and Intelligent Laboratory Systems*, *72* (1), 27–34.

[27] Chou, S.-M., Lee, T.-S., Shao, Y.E. & Chen, I.-F. (2004). Mining the breast cancer pattern using artificial neural networks and multivariate adaptive regression splines. *Expert Systems with Applications*, *27* (1), 133–142.

[28] de Cos Juez, F.J., Sánchez Lasheras, F., García Nieto, P.J. & Suárez Suárez, M. A. (2009). A new data mining methodology applied to the modelling of the influence of diet and lifestyle on the value of bone mineral density in post-menopausal women. *International Journal of Computer Mathematics*, *86* (10), 1878–1887.

[29] Vidoli, F. (2011). Evaluating the water sector in Italy through a two stage method using the conditional robust nonparametric frontier and multivariate adaptive regression splines. *European Journal of Operational Research, 212* (13), 583–595.

[30] García Nieto, P.J., Sánchez Lasheras, F., de Cos Juez, F.J. & Alonso Fernández, J.R. (2011). Study of cyanotoxins presence from experimental cyanobacteria concentrations using a new data mining methodology based on multivariate adaptive regression splines in Trasona reservoir (Northern Spain). *Journal of Hazardous Materials, 195*, 414–421.

[31] García Nieto, P.J., Alonso Fernández, J.R., Sánchez Lasheras, F., de Cos Juez, F.J. & Díaz Muñiz, C. (2012). A new improved study of cyanotoxins presence from experimental cyanobacteria concentrations in the Trasona reservoir (Northern Spain) using the MARS technique. *Science of Total Environment, 430*, 88-92.

[32] Colbeck, I. (2008). *Environmental Chemistry of Aerosol*. New York: Wiley-Blackwell.

[33] Hewitt, C.N. & Jackson, A.V. (2009). *Atmospheric Science for Environmental Scientists*. New York: Wiley-Blackwell.

[34] Schnelle, K.B. & Brown, C.A. (2001). *Air Pollution Control Technology Handbook*. Boca Ratón, Florida: CRC Press.

[35] Monteiro, A., Lopes, M., Miranda, A.I., Borrego, C. &Vautard, R. (2005). Air pollution forecast in Portugal: a demand from the new air quality framework directive. *International Journal of Environmental Pollution, 5*, 1–9.

[36] Friedlander, S.K. (2000). *Smoke, Dust and Haze: Fundamentals of Aerosol Dynamics*. New York: Oxford University Press.

[37] Vincent, J.H. (2007). *Aerosol Sampling: Science, Standards, Instrumentation and Applications*. New York: John Wiley & Sons.

[38] Anderson, W., Prescott, G.J., Packham, S., Mullins, J., Brookes, M. & Seaton, A. (2001). Asthma admissions and thunderstorms: a study of pollen, fungal spores, rainfall, and ozone. *Quarterly Journal of Medicine, 94* (8), 429–433.

[39] Weinhold, B. (2008). Ozone nation: EPA standard panned by the people. *Environmental Health Perspectives, 116* (7), A302-A305.

[40] Jerrett, M., Burnett, R.T., Arden Pope III, C., Ito, K., Thurston, G., Krewski, D., Shi, Y., Calle, E. &Thun, M. (2009). Long-term ozone exposure and mortality. *The New England Journal of Medicine, 360* (11), 1085–1095.

[41] Freedman, D., Pisani, R. &Purves, R. (2007).
[42] *Statistics*. New York: W.W. Norton & Company.
[43] Efron, B. &Tibshirani, R. (1997). Improvements on cross-validation: the .632 + bootstrap method. *Journal of the American Statistical Association*, *92* (438), 548–560.
[44] Picard, R. & Cook, D. (1984). Cross-validation of regression models. *Journal of the American Statistical Association*, *79* (387), 575–583.

In: Air Quality
Editor: Arthur Hermans

ISBN: 978-1-62808-259-3

Chapter 3

THE IMPACT OF AIR POLLUTION FROM THE MINING-METALLURGICAL COMPLEX ON THE CONTENT OF TOTAL SULFUR IN PLANT MATERIAL AND SOIL

Snezana M. Serbula[1,*], Tanja S. Kalinovic[1], Ana A. Ilic[1], Jelena V. Kalinovic[1] and Branko M. Bugarski[2]

[1]University of Belgrade, Technical Faculty in Bor, Bor, Serbia
[2]University of Belgrade, Faculty of Technology and Metallurgy, Beograd, Serbia

ABSTRACT

The area of Bor and the surroundings (Eastern Serbia) have been known for exploitation and processing of sulfide copper ores for more than a hundred years. Long-term ore exploitation in the examined area has made a great impact on local environment. For the purpose of evaluating the influence of emission from the mining-metallurgical complex, biomonitoring has been conducted in the urban-industrial and rural zone. Total sulfur concentrations were determined in samples of deciduous (birch and linden) and evergreen (spruce and pine) trees, as well as in moss and edible parts of fruits and vegetables (apple, pear, string beans and potato).

[*] E-mail: ssherbula@tf.bor.ac.rs. Tel.: +381 30 424 555; Fax: +381 30 421 078.

Significantly higher concentrations of the total sulfur in parts of woody plant species have been detected at the sampling sites Town Park (birch leaves 5745 μg g^{-1} dw) and Hospital (birch leaves 5668 μg g^{-1} dw) in the urban-industrial zone, which has been confirmed with the enrichment factors (EF). Comparing foliar parts, enrichment with sulfur decreases in the following order: birch leaves > linden leaves > spruce needles. Significantly higher concentrations of sulfur (and hence EF) in moss samples confirm the reliability of moss as a bioindicator of air pollution. In the sampled edible parts of fruits in the rural areas, substantially lower concentrations of the total sulfur have been detected (197.3-308.6 mg kg^{-1} dw) compared to vegetables (359.0-1119.4 mg kg^{-1} dw). Correlations among the sampling sites, based on the concentrations of the total sulfur in plant parts and soil of linden, birch, spruce and pine show the closest relationship among the Town Park, Hospital and Slatina ($R^2 \geq 0.8$), which indicates a similar level of pollution in the urban-industrial and rural zone.

Keywords: Passive biomonitoring, Air pollution, Total sulfur, Copper smelter

1. INTRODUCTION

The basic components of biosphere (vegetation, soil, air and water) are affected by atmosferic deposition of pollutants. The quality of air in a certain area depends mostly on anthropogenic sources of pollution, such as transport and industry (Al-Alawi and Mandiwana 2007; Fowler et al. 2009). Metallurgical production of non-ferrous metals is a significant source of heavy metals and waste gases (Boyd et al. 2009). Smelters, in which ores with a high content of sulfur are melted, emit SO_2 which is harmful to the environment in excessive concentrations (Derome and Lindroos 1998; Saarela et al. 2005; Koptsik and Alewell 2007). Secondary sources of environmental pollution in the vicinity of the mining and metallurgical plants represent tailings from which the particles are distributed by wind (Csavina et al. 2011). Due to the increased air pollution close to mining and metallurgical complexes, different types of monitoring are implemented (Šerbula et al. 2010; Serbula et al. 2012a; Serbula et al. 2013a; Serbula et al. 2013b).

Air quality monitoring using biological material has been in use for over a century. Initially, biomonitoring was applied only to analyze the impact of SO_2 from the air (as the main air pollutant) on biomaterial. At the present time, biomonitoring is used to monitor a number of pollutants, since it is a reliable way to make a connection between air pollution in large areas and the temporal and spatial concepts at an affordable price (Falla et al. 2000; Wolterbeek 2002). Aboveground parts of vegetation are the first level of deposition of atmosferic pollutants, and represent an aerosol filter. Diameter of most particles in the urban atmosfere is less than 0.1 μm, unlike the diameter of stomas (openings on the leaf surface), which is between 5 μm and 30 μm, so that the process of foliar absorption of pollutants from the air is undisturbed (Kadović and Knežević 2002; Jablanović et al. 2003; Milanovic et al. 2010; Milanovic et al. 2011).

As passive bioindicators of air pollution, the following species are the most commonly used: pine (Rautio et al. 1998; Rossini Oliva and Mingorance 2006; Sun et al. 2009; Lehndorff and Schwark 2010; Serbula et al. 2013b), spruce (Brunner et al. 2008; Stojakovic et al. 2012), birch (Kozlov 2005), linden (Dmuchowski and Bytnerowicz 2009; Tomašević et al. 2011), oak (Madejon et al. 2006), ash (Catinon et al. 2009), acacia (Samecka-Cymerman et al. 2009; Serbula et al. 2012a), lichens (Salemaa et al. 2004; Giordano et al. 2005), mosses (Poikolainen et al. 2004; Rivera et al. 2011; Tretiach et al. 2011). The most common use of mosses for bioindication of air pollution is justified for a number of reasons. Mosses have no root system, so the nutrients are absorbed from the air, their morphology does not vary seasonally, which means that the accumulation of pollutants takes place throughout the year. Also, they have a long life, so they are used for long-term monitoring of air pollution (Fernandez et al. 2002; Wolterbeek 2002; Dmuchowski and Bytnerowicz 2009).

Sulfur (S) is an important nutrient. Plants compensate their need for sulfur, absorbing it from the soil in the form of sulfate ions, and partly by foliar absorption from the air, which is observed especially in industrial areas (Jablanović et al. 2003). According to Cicek and Koparal (2004), the amount of sulfur in the leaves reflects deposition of SO_2 from the atmosfere. Discarded parts of plants (eg. leaf and fruit) increase the concentration of S in the soil because they are subject to decomposition and oxidation on the surface of the soil. During that process, sulfur gets transformed to a form acceptable to plants (Jablanović et al. 2003).

However, Sun et al. (2009) argue that the absorption of S from soil is well regulated and does not cause increased foliar concentrations. When the concentration of S in the soil is sufficient, excessive concentration of SO_2 in the air can have a toxic effect on vegetation (Yilmaz and Zengin 2004; EA 2008).

The natural level of SO_2 in the air (about 5 μg m^{-3}) is not toxic to plants (Jablanović et al. 2003). Interestingly, the concentrations of SO_2, NO_x and O_3, which are below the maximum allowable concentrations safe for humans, can cause serious damage to plants (WHO 2000). According to Jablanović et al. (2003), the concentrations of SO_2 <50 μg m^{-3} in the air may be useful as a source of the necessary S, while concentrations >100 μg m^{-3} cause adverse effects, even in tolerant plant species. According to the World Health Organization (WHO 2000), the critical concentration of SO_2 in the air, on an annual basis, for the forest and vegetation is 30 μg m^{-3}, while for agricultural crops it is 20 μg m^{-3}.

The aim of the survey was to assess which woody plant species (linden, birch, spruce, or pine) is the best passive bioindicator of sulfur air pollution in the vicinity of the copper smelter in the three studied zones. The concentration of total sulfur in the leaves/needles, branches and roots of the four plant species should indicate the areas with higher and lower levels of pollution than the dominant source of pollution.

For comparison, the concentration of total sulfur content was determined in the samples of moss, fruits (apples and pears) and vegetables (potatoes and green beans).

2. Methodology

2.1. The Study Area

The primary sources of environmental pollution in Bor (Serbia) are mining (surface mining of copper ore) and metallurgy (pyrometallurgical copper production from the following sulfide ores: chalcopyrite ($CuFeS_2$), chalcozine (Cu_2S) and covelline (CuS)) (Antonijevic et al. 2005a; Šerbula et al. 2010).

Additional sources of air pollution in Bor and surroundings are flotation tailings and open pits, from which particles are emitted with a high concentration of sulfates (Antonijevic et al. 2007; Antonijevic et al. 2008; EIA Study 2010). The Bor copper mine started working in 1903, with the discovery of rich reserves of copper. In 1906, the first copper smelter started its operations. In the period 1961-1968, a smelter was built where sulfide copper ore with accompanying elements (Zn, Pb, Ni, Au, Ag, As etc.) is melted even today.

Waste gases from the smelter have been emitted into the atmosfere without treatment until 1960. In order to protect the environment, in the same year sulfuric acid plant was built. Smelter waste gases containing 3-7% SO_2, are converted to H_2SO_4 in the plant. The plant provides accepting and processing less than 60% of the waste gases from the smelter, while the remainder is emitted into the atmosfere (Šerbula et al. 2010). For many years, copper smelter has been the dominant source of environmental pollution in Bor and beyond (SEPA 2003-2010).

The main pollutants are SO_2, particulate matter with a high content of As, Pb, Cu, Zn, Cd, and aero sediments, which also contain heavy metals and sulfates (Šerbula et al. 2011; Serbula et al. 2012b).

This type of pollution is highly corrosive, and it has been a subject of many researches (Serbula et al. 2002; Antonijevic et al. 2005b; Milic et al. 2008).

In Figure 1. a map of the study area is presented with the sites where sampling of plant species (linden, pine, spruce, birch, moss, fruits and vegetables) was carried out. The sites were selected in respect of copper smelter position within the mining-metallurgy complex and winds of the highest frequency (Figure 2.). Sulfur dioxide measuring sites are also shown on the map.

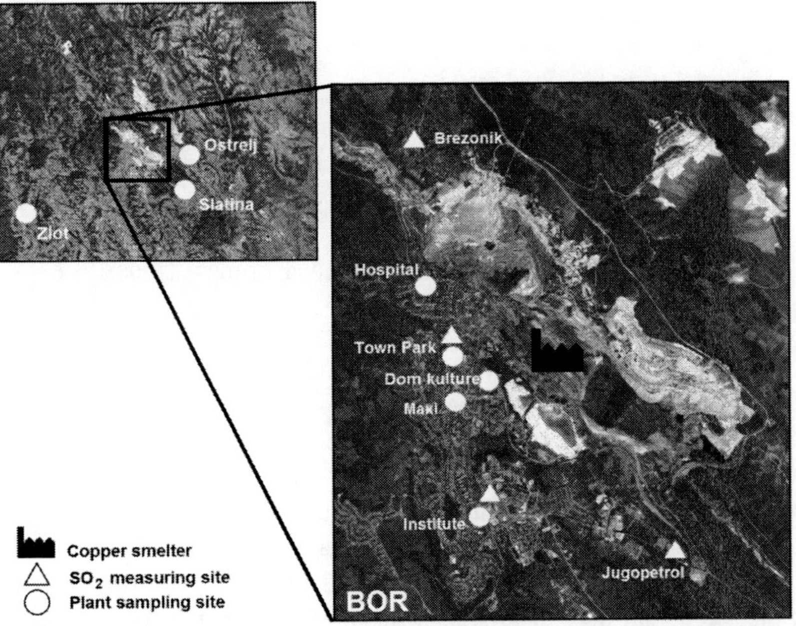

Figure 1. Map of Bor and its surroundings (Eastern Serbia) showing sampling sites of plant species and soil, sulfur dioxide measuring sites and position of the copper smelter.

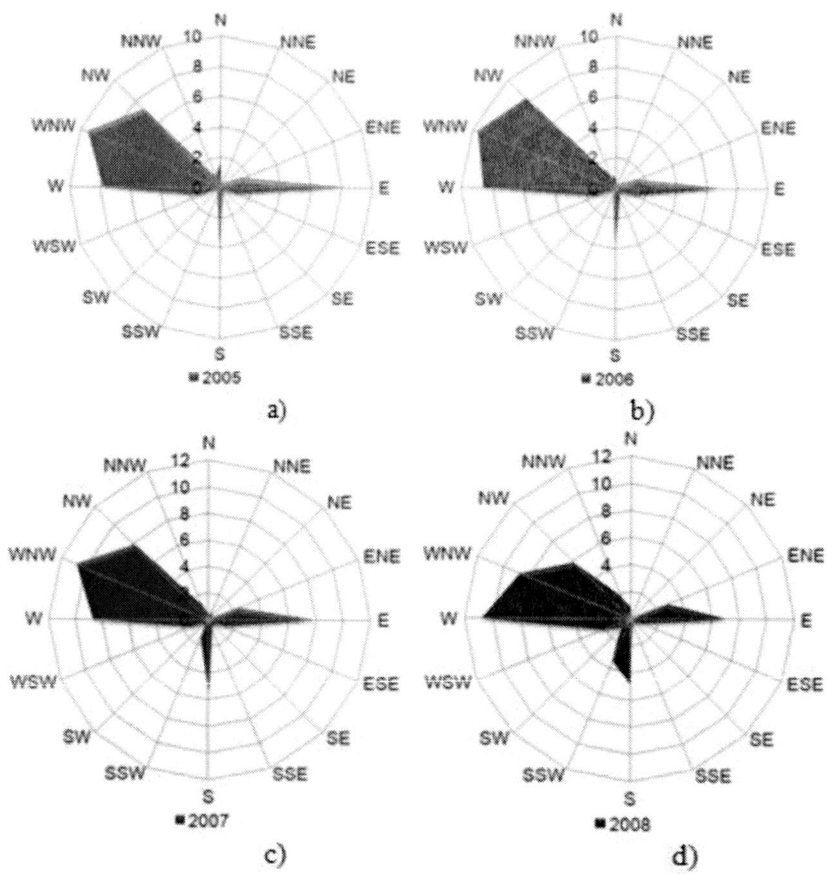

Figure 2. Windrose diagrams (%) for a) 2005; b) 2006; c) 2007; d) 2008.

2.1.1. Air Pollution Data

Pollution roses based on monthly SO_2 concentrations at the four measuring sites in the urban-industrial zone of Bor in the period from 2005 to 2008 (MMI 2005-2008) are given in Figure 3. At the measuring sites Town Park and Jugopetrol, SO_2 concentrations were higher compared to the concentrations at the sites Insitut and Brezonik. From Figure 3. it can be seen that the highest SO_2 concentration reached 450 $\mu g\ m^{-3}$ at the measuring site Jugopetrol in February 2005. Pollution roses for four measuring sites during 2005 and 2008 shows that months with the lowest SO_2 concentrations were August and September, as well as January. This could be a consequence of a reduced plant operation and/or favourable weather conditions.

During the 4-year period, SO_2 concentrations at the most polluted sites were above the EU limit value (LV), the Serbian LV and the US National Ambient Air Quality Standards (NAAQS) (EC 2008; The Official Gazette of The Republic of Serbia 2010; US EPA 2010). According to the detected exceedances of LVs it could be concluded that plants in Bor and its surroundings grow in the environment which is toxic for plants (WHO 2000).

Serbula et al. (2013a) showed that SO_2 concentrations mostly followed the anode copper production in the period from 2005 to 2008. In some cases, the trend was not present and increasing production was accompanied by decreasing SO_2 concentrations and vice versa, due to the irregularities during the operation of the sulfuric acid plant.

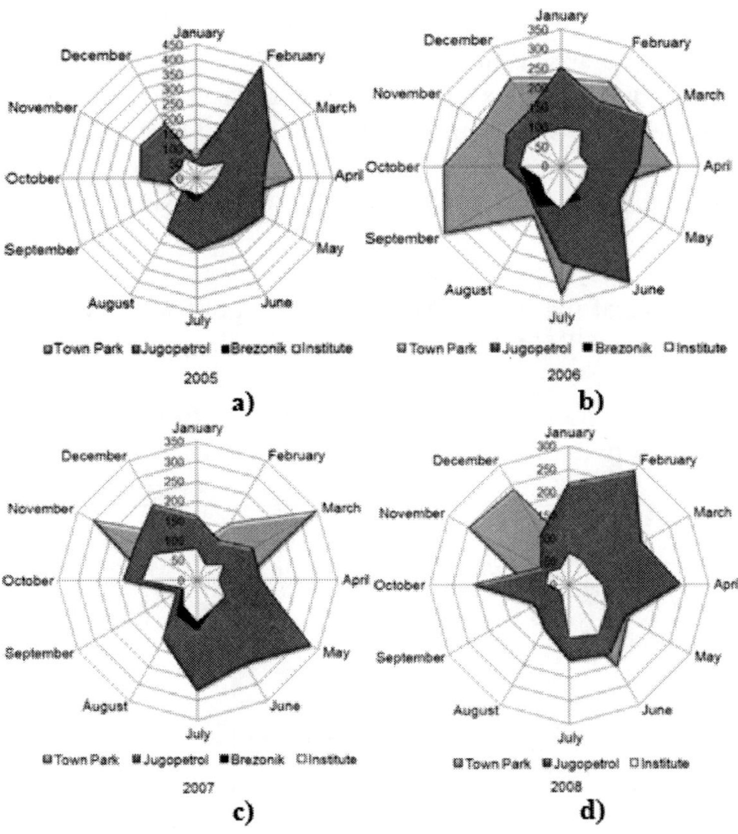

Figure 3. Pollution roses based on the monthly SO_2 concentrations ($\mu g\ m^{-3}$) at the measuring sites Town Park, Jugopetrol, Brezonik and Institute for a) 2005; b) 2006; c) 2007; d) 2008 (except the measuring site Brezonik).

This is the consequence of frequent stoppages of the smelter and difficulties in operation due to limited capacities of sulfuric acid plant.

2.2. Sampling of Biomaterial and Sample Preparation

Locations for sampling biomaterials were chosen in relation to: the level of air pollution, the position of the mining-metallurgical complex and wind direction with the highest frequency (Table 1, Figure 1.). The area of research is divided into three zones, two of which are polluted (urban-industrial and rural), and one background. Urban-industrial (UI) zone includes the sampling sites Town Park, Hospital, Dom kulture, Maxi, Institute. The rural (R) zone includes sampling sites in rural areas Slatina and Ostrelj, and the background (B) zone is a rural village Zlot. Leaves/needles, branches and roots were sampled from three trees per each plant species. The samples of branches with leaves/needles were sampled from a height of about 2 m above the ground. For a more complete analysis, the soil was sampled under the canopy of each tree, and so were the roots (first washed with running water and then with distilled water).

First, the organic surface layers of the soil were removed (0-5 cm), and then about 2 kg of subsurface layer was collected of 5 to 20 cm in depth. The soil samples were sifted through a stainless steel sieve, mesh size 0.883 mm.

Table 1. Sampling sites, wind directions, distance from the copper smelter and availability of woody, fruit and vegetable plant species

Sampling site	Wind direction[a]	Distance[b] (km)	Plant species							
			pine	spruce	linden	birch	apple	pear	potato	string beans
Town park	ENE	0.5	-	+	+	+	-	-	-	-
Hospital	ESE	1.0	-	-	+	+	-	-	-	-
Ostrelj	WNW	4.5	+	+	+	+	+	+	+	+
Slatina	NW	6.5	+	+	+	+	+	+	+	+
Zlot	NE	13.0	+	+	+	+	+	+	+	+

[a]wind that brings pollution from the direction of copper smelter.

[b]distance of the sampling site from the copper smelter.

+plant species were sampled.

-plant species were not sampled.

From the sifted soil samples, after homogenization, a representative subsample was taken, weighing 100±1 g, which was finely ground in a laboratory mill to a particle size of 100 μm. All the samples were dried at room temperature for 10 days, and then in a dryer at a temperature of 50°C for more than 24 hours. Each specimen of plant material (leaves/needles, branches, roots) and soils represent a homogeneous sample from a particular sampling site. In the same period, moss was sampled at four sites in the urban-industrial area and the background sampling site (Figure 1.): Town Park, Dom kulture (0.6 km to the W), Maxi (0.9 km to the WSW), Institute (2.5 km to the SSW) and Zlot. Moss sampling was done independently of the sampling of woody plants, due to the different availability of species. Moss samples were prepared in the same manner as the plant material of woody species.

The edible parts of fruits and vegetables were sampled from a few spots at the sites Ostrelj, Slatina and Zlot (Table 1, Figure 1.). Fresh samples were washed in a laboratory with running water first and then with distilled water. After that they were dried at room temperature for 10 days and then in a dryer at 50°C for more than 24 hours, after which they were ground in a laboratory mill to a particle size of 100 μm. Each sample of the edible parts of fruits and vegetables is a representative homogeneous sample from a particular sampling site. Digestion of all the samples was performed by the method of the U.S. Environmental Protection Agency 3050B (US EPA 1996). The prepared samples in quantity of 0.1000 g dry mass were digested in the mixture of nitric acid (HNO_3, 65%, *p.a.)* and hydrogen peroxide (H_2O_2, 30%) in a volume ratio of 4:1, as in the papers of Husain et al. (1995), Piczak et al. (2003), Rossini Oliva and Mingorance (2006) and Lindroos et al. (2007).

Determination of total sulfur concentration in soil samples and biomaterial was performed at the Institute of Mining and Metallurgy in Bor, by inductively coupled plasma atomic emission spectrometry (ICP-AES, model "Spectro Ciros Vision"). Lower limit of detection for sulfur was 25 μg g^{-1}. The concentration of total sulfur is shown in units of μg g^{-1} dry weight.

3. Results and Discussion

3.1. Concentrations of Total S in Woody Plant Species

The determined total S concentrations in woody species were aimed to indicate the appropriate plant part for the biomonitoring of sulfur on the territory of Bor and its surroundings.

Concentrations of total S in the leaves/needles, branches and roots of linden, birch, spruce and pine, as well as soil at the sampling sites in the urban-industrial, rural and background zone are given in Table 2.

Table 2. Total S concentrations (μg g^{-1} dw) in plant parts and soil of linden, birch, spruce and pine

Species	Plant part/soil	Sampling sites				
		Town Park	Hospital	Ostrelj	Slatina	Zlot
Linden	Leaves	4773	4555	1874	2616	2469
	Branches	1429	838	467	817	412
	Roots	1553	871	731	779	581
	Soil	758	358	580	400	384
Birch	Leaves	5745	5668	1715	2091	1394
	Branches	674	554	351	414	521
	Roots	465	480	466	937	989
	Soil	806	1017	1620	868	473
Spruce	Needles	2453	/	903	1863	1766
	Branches	1137	/	675	1039	265
	Roots	1158	/	756	552	708
	Soil	982	/	2058	676	1205
Pine	Needles	/	/	1276	1574	831
	Branches	/	/	668	1017	312
	Roots	/	/	1030	2339	880
	Soil	/	/	663	1324	184

/ not sampled

Maximum concentrations of S in all parts of linden and topsoil were detected in the urban-industrial area. Concentrations of S in leaves, branches and roots of linden decrease with the distance of the sampling site from the copper smelter, except for the sampling site Slatina and Ostrelj. Between the sampling site Ostrelj and the dominant source of pollution there is a physical barrier (the hill) during transport of pollutants (Serbula et al. 2013b). In the conditions of lower wind speed, the topography of the terrain has a significant impact on air pollution, and thus on the S concentration in the plant material. Therefore, at the sampling site Ostrelj the concentrations of S are lower than at the sampling site Slatina. Based on the concentrations of total S (Table 2), for all the sampling sites (except for Slatina), a descending order can be formed in parts of linden: leaves>roots>branches. For soil samples, a similar tendency of decreasing S concentrations with the increasing distance from the dominant source of pollution was not noticed. In general, lower concentrations of S were detected in soil compared to the S content in plant material of linden.

In the leaves and soil of birch, the maximum concentrations of total S were detected at the sampling sites Town Park, Hospital and Ostrelj, while at the sampling sites Slatina and Zlot the maximum concentrations were detected in leaves and roots (Table 2). S concentrations in leaves of birch and linden decreases with the distance of sampling from the copper smelter, except at the sites Slatina and Ostrelj, which is due to the local topography. Reimann et al. (2001) argue that birch leaves do not indicate air pollution by sulfur dioxide, unlike Hrdlicka and Kula (2004), who found that the birch responded to the reduction of air pollution by sulfur dioxide. Reimann et al. (2001) have detected S concentration of 1775 $\mu g\ g^{-1}$ in unwashed leaves of birch sampled near the Ni smelter, and 1800 $\mu g\ g^{-1}$ in an area with urban pollution (Reimann et al. 2007). Steinnes et al. (2000) have detected S in concentration of 4270 $\mu g\ g^{-1}$ in birch unwashed leaves at the sampling sites with increased pollution, which is a lower value than the value of this research.

At the sampling site nearest to the dominant source of pollution in the study area (Town Park), concentrations of total S in needles, branches and roots of spruce are the highest, compared to the other sampling sites. Sun et al. (2009) found that concentrations in current year's and the previous year's needles were over 1200 $\mu g\ g^{-1}$ and 1500 $\mu g\ g^{-1}$ dw, respectively, indicating air pollution with sulfur. According to Petkovšek et al. (2008), spruce needles are ranked into four classes based on the content of total S. It can be concluded that spruce needles sampled in Bor and the surroundings, belong to the fourth class with the highest S content (>1.58 $\mu g\ g^{-1}$). At the sampling site Ostrelj, the highest concentration of S was detected in the soil of spruce, which may be a result of the impact of sulfate particles emitted from tailings located near the village. S concentration of 1985 $\mu g\ g^{-1}$ in unwashed spruce needles was a consequence of the impact of pollution from the Ni smelter (Reimann et al. 2001), and the value of 900 $\mu g\ g^{-1}$ was detected in the urban atmosfere.

Higher concentrations of total S in parts of pine, as well as in the soil were detected in the rural areas compared to the background area (pine was not sampled in the urban-industrial area). At the sampling site Slatina higher concentrations were detected in plant material and soil compared to the sampling site Ostrelj (effects of terrain topography). S concentration in the pine at the sampling sites Slatina and Zlot declined in the following order: roots>needles>branches, while the order of S at the sampling site Ostrelj was: needles>roots>branches. Therefore, roots and needles contained S in the highest concentrations. Normal concentration range of sulfur in the pine needles is from 500 μg g^{-1} to 850 μg g^{-1} (Manninen et al. 1996; Rautio et al. 1998), while the critical concentration is 900 μg g^{-1} (Manninen et al. 1996). Based on the concentrations in Table 2, it can be concluded that S in pine needles is in the range of normal concentrations only at the background site. At all the other sampling sites, concentrations are higher, which suggests a direct influence of SO_2 emissions from the smelter. The concentrations of S in unwashed pine needles near the Cu-Ni smelter ranged from 678 μg g^{-1} to 1941 μg g^{-1} (Rautio i sar. 1998), while in the vicinity of the Ni smelter they were 1190 μg g^{-1} (Reimann et al. 2001). In the urban area of Cologne (Germany), Lehndorff and Schwark (2010) detected S in unwashed pine needles in the range of 868 μg g^{-1} to 2076 μg g^{-1}. In the paper Sun et al. (2009), the highest S concentration in washed pine needles was detected at the sampling site in the industrial area (3080 μg g^{-1}), and the lowest concentration was at the sampling site in a natural reserve, which indicates a direct influence of anthropogenic pollution. Based on the data from Table 2, there is a regular decrease of S concentration in the pine soil in the following order: Slatina>Ostrelj>Zlot, ie. concentration decreases with distance from the source of pollution. In the paper Koptsik and Alewell (2007), S concentrations in soil were also higher at the sampling sites closer to Cu-Ni smelter. Sun et al. (2009), detected S in the amount of about 300 μg g^{-1} in the pine soil (0-10 cm) in the industrial area, which is a lower value than the concentration in the rural areas in this survey.

3.2. Concentrations of Total S in Moss

Fine-scale moss monitoring technique is an efficient method for the definition of small and large sources of pollution, as well as for the quantitative determination of atmosferic deposition of pollutants (Sucharova and Suchara 2004).

The obtained results from our survey are in agreement with the numerous studies (Sucharova and Suchara 2004; Krommer et al. 2007), according to which mosses are efficient bioindicator of air pollution by sulfur dioxide and certain types of moss can survive high levels of pollution, contrary to the conclusions of Reimann et al. (2001). The following concentrations of total S were detected: 7363 μg g^{-1} (at the sampling site Town Park), 7009 μg g^{-1} (Dom kulture), 7196 μg g^{-1} (Maxi), 3674 μg g^{-1} (Institute) and 1482 μg g^{-1} (Zlot). The concentration of total S in moss, sampled in the urban and industrial areas (Town Park, Dom kulture, Maxi) is significantly higher than at the site Zlot and at the site which is not in the direction of the dominant winds (Institute).

According to Sucharova and Suchara (2004), a typical S content in moss is in the range 1200-1400 μg g^{-1}. This range of concentrations includes only S concentrations in moss sampled in the background zone, while the S concentration in the moss at the other locations is well above the range of typical concentrations. From the literature data (Reimann et al. 2001; Sucharova and Suchara 2004; Krommer et al. 2007), it can be concluded that the total S concentrations in moss samples are significantly lower than the values reported in this survey.

3.3. The Concentrations of Total S in Fruits and Vegetables

Figure 4. represents a box plot diagram of the total S concentration in edible parts of apples, pears, potatoes and green beans, sampled in the rural and background zone. The contents of total S in all the samples of fruits and vegetables varied in the range of 197.3 mg kg^{-1} to 1119.4 mg kg^{-1}, with higher concentrations being detected in vegetables. There are no available data on the concentrations of total S in fruits and vegetables, so the level of contamination of fruit can not be determined with certainty in the area of our research.

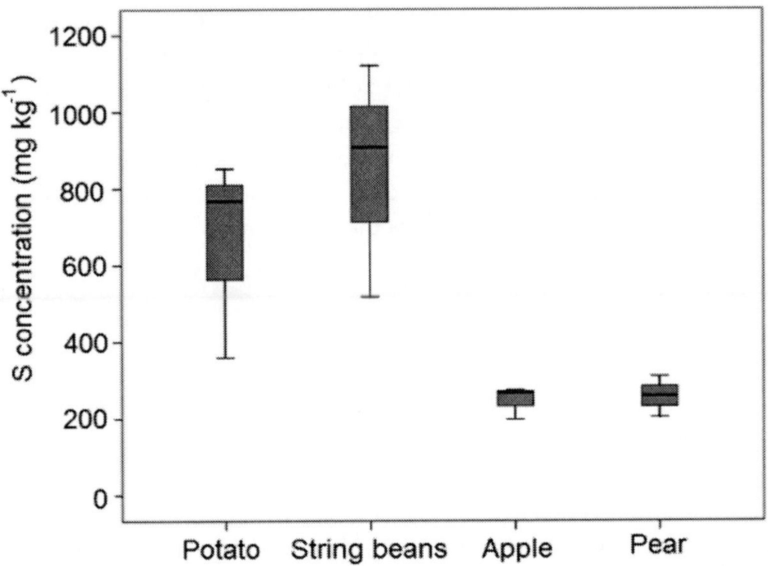

Figure 4. Total S concentrations in edible parts of vegetables and fruits sampled in the Bor surroundings.

However, in Hamurcu et al. (2010), the concentration of S in edible parts of apples growing at the roadside was 1449.64±60.97 mg kg^{-1}, which is several times higher than in our study area.

In addition to pollutants originating from copper smelter, fugitive dust from tailing ponds and uncontrolled use of agricultural chemicals can also cause an increase in total sulfur content in vegetables and fruits.

3.4. Correlations Among the Sampling Sites

Figure 5. shows the correlations among the sampling sites of woody plant species in Bor and its surroundings based on the total S concentrations in parts (roots, branches, leaves/needles) and soil of linden, pine, spruce and birch. Correlations with the highest value of the determination coefficient (R^2) are Town Park-Hospital (R^2>0.9), Hospital-Slatina (R^2>0.8) and Town Park-Slatina (R^2≈0.8).

The sampling site Town Park is located in the urban-industrial area which is characterized by the highest air pollution by sulfur dioxide (Figure 3.) due to the immediate proximity of the dominant source of pollution. The Hospital site belongs to the same urban-industrial zone as the Town park, which is a reason of very strong correlation (Figure 5a). Rural settlements, Slatina and Ostrelj are in direction of the dominant winds (Figure 2.). However, in plant material and soil, higher concentrations of total sulfur were detected at the sampling site Slatina compared to the site Ostrelj, which is why a stronger correlation was established with the sampling site Town Park (Figure 5b). As already mentioned, between the pollution source and the sampling site Ostrelj there is a barrier which to some extent, prevents the transport of pollutants to this site.

3.5. Enrichment Factors

The plant and soil enrichment factor (EF) has been calculated in order to derive the degree of contamination of plants growing at the polluted sites with respect to plants growing at the background site.

The EFs were calculated as $EF=C_{polluted}/C_{backgound}$, where $C_{polluted}$ and $C_{backgound}$ are total S concentrations (μg g^{-1}) in plant parts (leaves/needles, branches and roots) from the polluted sampling sites and the background site, respectively. According to Singh et al. (2010), the EF>1 is a measure of environmental pollution. Based on the enrichment factors given in Figure 6., it can be concluded that the higher level of SO_2 in the air has a significant effect on the total sulfur content in the examined plants and soil.

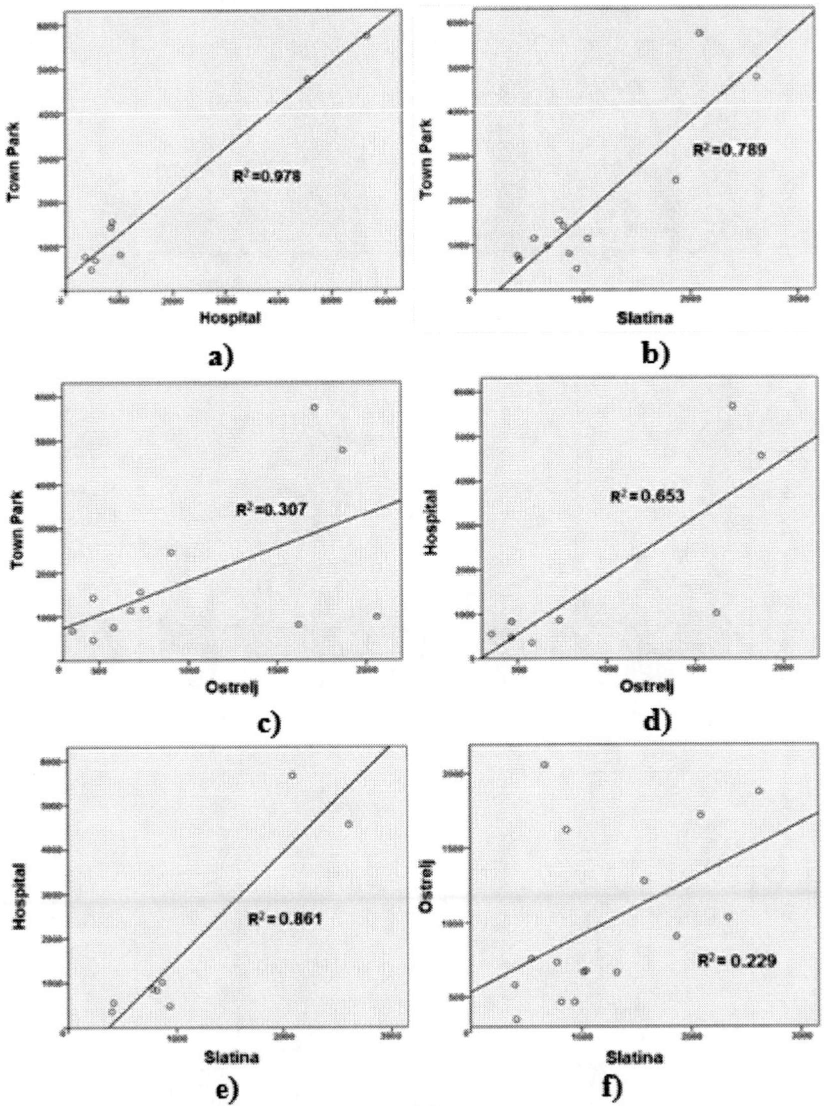

Figure 5. Correlations among the sampling sites based on the total S concentrations in plant parts and soil of four woody species: a) Town park-Hospital ($p<0.01$); b) Town park-Slatina ($p<0.01$); c) Town park-Ostrelj ($p<0.05$); d) Hospital-Ostrelj ($p<0.01$); e) Hospital-Slatina ($p<0.01$); f) Ostrelj-Slatina ($p<0.05$).

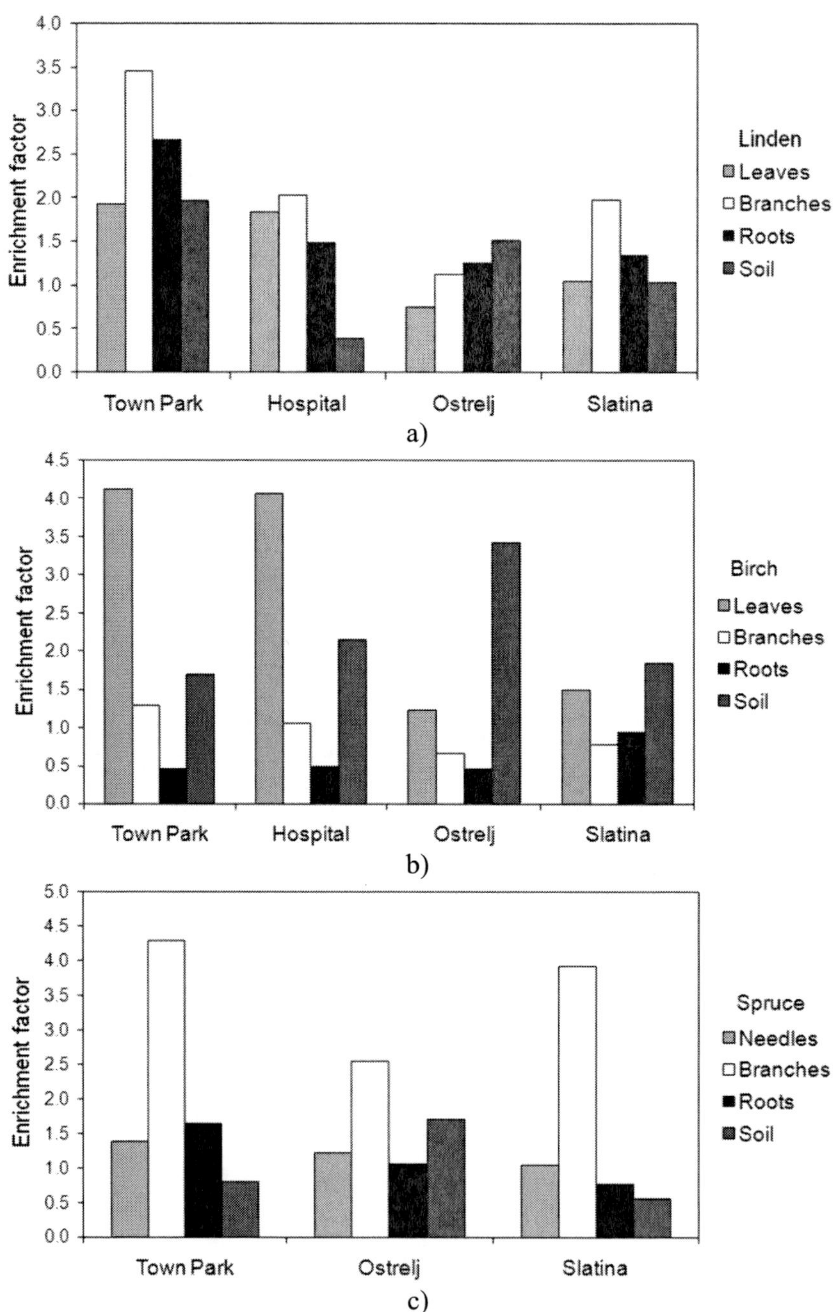

Figure 6. (Continued).

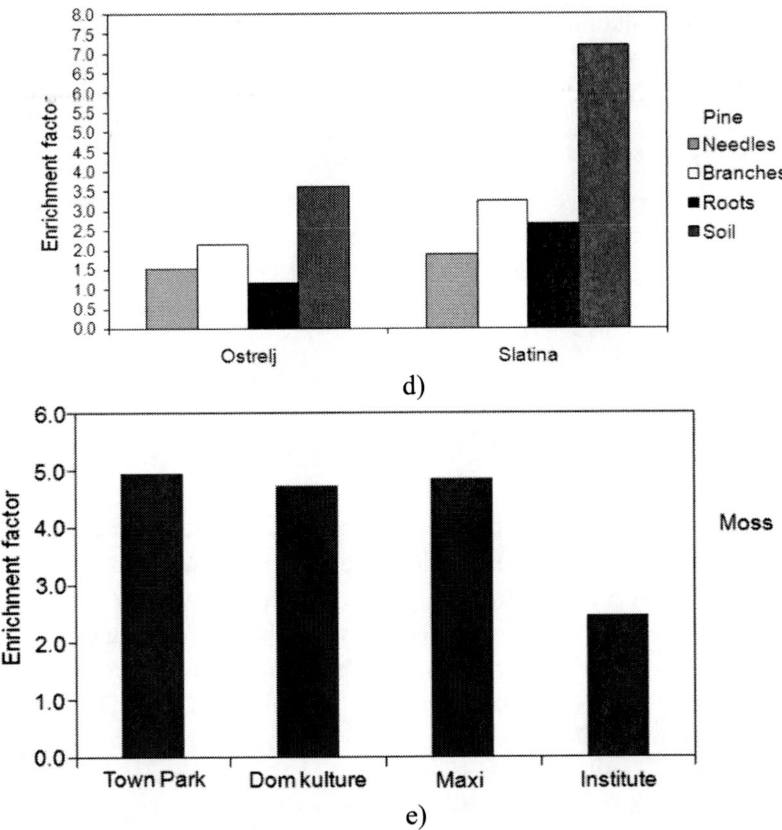

Figure 6. Plant and soil enrichment factors for a) linden; b) birch; c) spruce; d) pine; e) moss.

In all the samples of linden in the urban-industrial zone (Figure 6a) enrichment with sulfur is present, except in the soil at the sampling site Hospital. In the rural areas, the EFs were close to 1, except in the soil (Ostrelj) and branch (Slatina) of linden. The highest total sulfur enrichment (EF≥2) was in branches of linden at the sampling sites Town Park, Hospital and Slatina. In the samples of birch in the urban-industrial zone the enrichment with sulfur decreases in the following order leaves>soil>branches, while in the rural areas, enrichment is higher in soil than in leaves (Figure 6b). At the sampling sites of spruce the Town Park, Ostrelj and Slatina (Figure 6c) the highest enrichment was noted in the branches (EF>2.5), although the needles can also be used for biomonitoring of air pollution, considering that EF>1 at all the three sampling sites. Although pine was sampled only in the rural areas, it is obvious that pine responds to air pollution with sulfur compounds, because EF>1 in all the samples of plant material and soil. Moss sampled in the urban-industrial area (Figure 6e) has higher values of enrichment (EF>4.7) at the sites located from 0.5 to 0.9 km from the copper smelter compared to the site which is 2.5 km away from it (EF=2.48).

Considering the fact that foliar parts of woody plant species are mostly used for biomonitoring, in the conditions of the increased air pollution, the greatest enrichment was detected in birch leaves (EF≈4.1), then in linden leaves (EF<2), while spruce needles are the least enriched with sulfur (EF<1.5).

CONCLUSION

In order to assess the impact of emissions of pollutants from the mining-metallurgical complex on the environment of Bor and the surroundings, the analysis of the total S concentrations in two deciduous, two evergreen plant species, moss and edible parts of fruits and vegetables was conducted. At all the sampling sites in the leaves of deciduous trees, higher concentrations of total S were detected in relation to the branches, roots and soil. This suggests that linden and birch leaves provide a good response to air pollution, compared to spruce and pine needles, so they can be used as biomonitors of air pollution. As for spruce and pine, it can not be said with certainty which part is best for biomonitoring of air pollution, since no regular decrease of S concentration in plant parts was determined. Also, evergreen plant species were not sampled at all locations as deciduous. It was found that the branches of evergreen species are better indicators of air pollution in relation to deciduous ones. A similar regularity could not be determined for the roots. In soil samples, the lowest concentrations of total S were detected, which indicates that the sulfur in the sampled plant species originates, most likely, from foliar absorption from the air. It can not be stated with certainty about the impact of SO_2 emissions from the copper smelter on the level of the total S in the edible parts of fruits and vegetables due to possible uncontrolled use of agricultural chemicals. When comparing the concentrations of total S, moss proved to be a better bioindicator of air pollution than the examined woody plant species.

The analysis of the total S concentration at all the sampling sites in all parts of linden, birch, pine, spruce and the soil, it was found that the urban-industrial is the most polluted zone which includes the sampling sites Town

Park and Hospital. These sites are closest to the copper smelter, being at the dominant W and E wind direction.

Enrichment factors indicate that the increased levels of SO_2 in the air in the studied area has a significant impact on the total sulfur content in the sampled plants. Since foliar parts of woody plant species are mostly used for biomonitoring, in the conditions of the increased air pollution the greatest enrichment was detected in leaves of birch, then in linden leaves, while the lowest sulfur enrichment was detected in spruce needles. Moss sampled in the urban-industrial zone has higher enrichment compared to the parts of woody plant species.

High concentrations of total S in plant material from the area of research indicate its anthropogenic origin.

ACKNOWLEDGMENT

The authors are grateful to the Ministry of Education and Science of the Republic of Serbia for financial support (Projects No. 46010 and No. 33038).

REFERENCES

Al-Alawi, M. M. and Mandiwana, K. L. (2007). The use of Aleppo pine needles as a bio-monitor of heavy metals in the atmosfere. *Journal of Hazardous Materials,* 148 (1–2), 43–46.

Al Sayegh Petkovšek, S.; Batič, F. and Lasnik, C. R. (2008). Norway spruce needles as bioindicator of air pollution in the area of influence of the Šoštanj Thermal Power Plant, Slovenia. *Environmental Pollution*, 151 (2), 287–291.

Antonijevic, M. M.; Dimitrijevic, M. D.; Serbula, S. M.; Dimitrijevic, V. Lj.; Bogdanovic, G. D. and Milic, S. M. (2005a). Influence of inorganic anions on electrochemical behaviour of pyrite. *Electrochimica Acta,* 50, 4160–4167.

Antonijevic, M. M.; Milic, S. M.; Serbula, S. M. and Bogdanovic, G. D. (2005b). The influence of chloride ions and benzotriazole on the corrosion behavior of Cu37Zn brass in alkaline medium. *Electrochimica Acta,* 50, 3693–3701.

Antonijevic, M. M; Bogdanovic, G. D.; Serbula, S. M. and Milic, S. M. (2007). Influence of grain size on chalcopyrite ore leaching in acidic medium. *Journal of The Serbian Chemical Society*, 72 (8–9), 911–919.

Antonijevic, M. M.; Dimitrijevic, M. D.; Stevanovic, Z. O.; Serbula S. M. and Bogdanovic G. D. (2008). Investigation of the possibility of copper recovery from the flotation tailings by acid leaching. *Journal of Hazardous Materials,* 158, 23–34.

Boyd, R.; Barnes, S. J.; De Caritat, P.; Chekushin, V. A.; Melezhik, V. A., Reimann, C. and Zientek, M. L. (2009). Emissions from the copper–nickel industry on the Kola Peninsula and at Noril'sk, Russia. *Atmospheric Environment,* 43 (7), 1474–1480.

Brunner, I.; Luster J.; Günthardt-Goerg, M. S. and Frey, B. (2008). Heavy metal accumulation and phytostabilisation potential of tree fine roots in a contaminated soil. *Environmental Pollution*, 152 (3), 559–568.

Catinon, M.; Ayrault, S.; Clocchiatti, R.; Boudouma, O.; Asta, J.; Tissut, M. and Ravane, P. (2009). The anthropogenic atmospheric elements fraction: A new interpretation of elemental deposits on tree barks. *Atmospheric Environment*, 43 (5), 1124–1130.

Cicek, A. and Koparal, A. S. (2004). Accumulation of sulfur and heavy metals in soil and tree leaves sampled from the surroundings of Tuncbilek Thermal Power Plant. *Chemosphere,* 57 (8), 1031–1036.

Csavina, J.; Landázuri, A.; Wonaschütz, A.; Rine, K.; Rheinheimer, P.; Barbaris, B.; Conant, W.; Sáez, A. E. and Betterton, E. A. (2011). Metal and metalloid contaminants in atmospheric aerosols from mining operations. *Water, Air, and Soil Pollution*, 221 (1–4), 145–157.

Derome, J. and Lindroos, A. J. (1998). Effects of heavy metal contamination on macronutrient availability and acidification parameters in forest soil in the vicinity of the Harjavalta Cu-Ni smelter, SW Finland. *Environmental Pollution,* 99 (2), 225–232.

Dmuchowski, W. and Bytnerowicz, A. (2009). Long-term (1992–2004) record of lead, cadmium, and zinc air contamination in Warsaw, Poland: determination by chemical analysis of moss bags and leaves of Crimean linden. *Environmental Pollution*, 157 (12) 3413–3421.

EA (Environment agency), (2008). Review and implementation study of biomonitoring for assessment of air quality outcomes. Available from: http://a0768b4a8a31e106d8b0-50dc802554eb38a24458b98ff72d550b.r19.cf3.rackcdn.com/scho0907bngy-e-e.pdf

EC Directive (2008). Directive 2008/50/EC of the European Parliament and of the Council of 21 May 2008 on Ambient Air Quality and Cleaner Air for Europe, *Official Journal of the European Union*, L152, 11 June 2008, p. 1–44.

EIA Study-New (2010) Smelter and sulphuric acid plant project. Faculty of Metallurgy, University of Belgrade, SNC Lavalin, pp 1–16.

Falla, J.; Laval-Gilly, P.; Henryon, M.; Morlot, D. and Ferard, J.-F. (2000). Biological air quality monitoring: A review. *Environmental Monitoring and Assessment,* 64 (3), 627–644.

Fernandez, J. A.; Ederra, A.; Nunez, E.; Martinez-Abaigar, J.; Infante, M.; Heras, P.; Elias, M. J.; Mazimpaka, V. and Carballeira, A. (2002). Biomonitoring of metal deposition in northern Spain by moss analysis. *Science of The Total Environment*, 300 (1–3), 115–127.

Fowler, D.; Pilegaard, K.; Sutton, M. A.; Ambus, P.; Raivonen, M.; Duyzer, J.; Simpson, D.; Fagerli, H.; Fuzzi, S.; Schjoerring, J. K.; Granier, C., Neftel, A.; Isaksen, I. S. A., Laj, P.; Maione, M.; Monks, P. S.; Burkhardt, J.; Daemmgen, U.; Neirynck, J.; Personne, E.; Wichink-Kruit, R.; Butterbach-Bahl, K.; Flechard, C.; Tuovinen, J. P.; Coyle, M.; Gerosa, G.; Loubet, B., Altimir, N.; Gruenhage, L., Ammannl, C.; Cieslik, S.; Paoletti, E.; Mikkelsen, T. N.; Ro-Poulsen, H.; Cellier, P.; Cape, J. N.; Horvath, L.; Loreto, F.; Niinemets, U.; Palmer, P. I.; Rinne, J.; Misztal, P.; Nemitz, E.; Nilsson, D.; Pryor, S.; Gallagher, M.W.; Vesala, T.; Skiba, U.; Bruggemann, N.; Zechmeister-Boltenstern, S.; Williams, J.; O'Dowd, C.; Facchini, M. C.; de Leeuw, G.; Flossman, A.; Chaumerliac, N. and Erisman, J. W. (2009). Atmospheric composition change: Ecosystems–Atmosphere interactions. *Atmospheric Environment*, 43 (33), 5193–5267.

Giordano, S.; Adamo, P.; Sorbo, S. and Vingiani, S. (2005). Atmospheric trace metal pollution in the Naples urban area based on results from moss and lichen bags. *Environmental Pollution*, 136 (3), 431–442.

Hamurcu, M.; Ozcan, M. M.; Dursun, N. and Gezgi, S. (2010). Mineral and heavy metal levels of some fruits grown at the roadsides. *Food and Chemical Toxicology*, 48 (6), 1767–1770.

Hrdlicka, P. and Kula, E. (2004). Changes in the chemical content of birch (*Betula pendula Roth*) leaves in the air polluted Krusne hory mountains. *Trees*, 18 (2), 237–244.

Husain, A.; Baroon, Z.; AI-khalafawi, M.; AI-Ati, T. and Sawaya, W. (1995). Toxic metals in imported fruits and vegetables marketed in Kuwait. *Environment International*, 21 (6), 803-805.

Jablanović, M. and Jakšić, P., Kosanović, K. (2003). Uvod u ekotoksikologiju. Beograd, Univerzitet u Prištini, Kosovska Mitrovica, Prirodno-matematički fakultet, HELETA.

Kadović, R., and Knežević, M. (2002). Teški metali u šumskim ekosistemima Srbije. Beograd, FINEGRAF.

Koptsik, G. and Alewell, C. (2007). Sulphur behaviour in forest soils near the largest SO_2 emitter in northern Europe. *Applied Geochemistry*, 22 (6), 1095–1104.

Kozlov, M. V. (2005). Sources of variation in concentrations of nickel and copper in mountain birch foliage near a nickel-copper smelter at Monchegorsk, north-western Russia: results of long-term monitoring. *Environmental Pollution,* 135 (1), 91–99.

Krommer, V.; Zechmeister, H. G., Roder, I.; Scharf, S. and Hanus-Illnar, A. (2007). Monitoring atmospheric pollutants in the biosphere reserve Wienerwald by a combined approach of biomonitoring methods and technical measurements. *Chemosphere,* 67 (10), 1956–1966.

Lehndorff, E. and Schwark, L. (2010). Biomonitoring of air quality in the Cologne Conurbation using pine needles as a passive sampler-Part III: Major and trace elements. *Atmospheric Environment*, 44 (24), 2822–2829.

Lindroos, A.; Derome, J.; Raitio, H. and Rautio P. (2007). Heavy metal concentrations in soil solution, soil and needles in a Norway Spruce stand on an acid sulphate forest soil. *Water, Air, and Soil Pollution*, 180 (1–4), 155–170.

Madejon, P.; Maranon, T. and Murillo, J. M. (2006). Biomonitoring of trace elements in the leaves and fruits of wild olive and holm oak trees. *Science of The Total Environment,* 355 (1–3), 187–203.

Manninen, S.; Huttunen, S.; Rautio, P. and Peramaki P. (1996). Assessing the critical level of SO_2 for scots pine in situ. *Environmental Pollution,* 93 (1), 27-38.

Milanovic, J.; Manojlovic, V.; Levic, S.; Rajic, N.; Nedovic, V. and Bugarski, B. (2010). Microencapsulation of flavours in Carnauba Wax. *Sensors*, 10 (1), 901–912.

Milanovic, J.; Levic, S.; Manojlovic, V.; Nedovic, V. and Bugarski, B. (2011). Carnauba wax microparticles produced by melt dispersion technique. *Chemical Papers,* 65 (2), 213–220.

Milic, S. M.; Antonijevic, M. M.; Serbula, S. M. and Bogdanovic, G. D. (2008). The Influence of Benzotriazole on the Corrosion Behaviour of CuAlNiSi Alloy in Alkaline Medium. *Corrosion Engineering Science and Technology,* 43, 30–37.

MMI (Mining and Metallurgy Institute Bor) (2005-2008). Reports on quality of ambient air in Bor from 2005 to 2008, Sector for measuring and control of gaseous and dust parameters, Department for Chemical-Technical Control.

Piczak, K.; Lesniewicz, A. and Zyrnicki, W. (2003). Metal concentrations in deciduous leaves from urban areas in Poland. *Environmental Monitoring and Assessment,* 86 (3), 273–287.

Poikolainen, J.; Kubin, E.; Piispanen, J. and Karhu, J. (2004). Atmospheric heavy metal deposition in Finland during 1985–2000 using mosses as bioindicators. *Science of The Total Environment,* 318 (1–3), 171–185.

Rautio, P.; Huttunen, S. and Lamppu, J. (1998). Element concentrations in scots pine needles on radial transects across a subarctic area. *Water, Air, and Soil Pollution,* 102, 389–405.

Reimann, C.; Koller, F.; Kashulina, G.; Niskavaara, H. and Englmaier, P. (2001). Influence of extreme pollution on the inorganic chemical composition of some plants. *Environmental Pollution,* 115 (2), 239–252.

Reimann, C.; Arnoldussen, A.; Boyd, R.; Finne, T. E.; Koller, F.; Nordgulen, O. and Englmaie, P. (2007). Element contents in leaves of four plant species (birch, mountain ash, fern and spruce) along anthropogenic and geogenic concentration gradients. *Science of The Total Environment*, 377 (2–3), 416–433.

Rivera, M.; Zechmeister, H.; Medina-Ramon M.; Basagana, X.; Foraster, M.; Bouso, L.; Moreno, T.; Solanas, P.; Ramos, R.; Kollensperger, G.; Deltell, A.; Vizcaya, D. and Kunzli N. (2011). Monitoring of heavy metal concentrations in home outdoor air using moss bags. *Environmental Pollution,* 159 (4), 954–962.

Rossini Oliva, S. and Mingorance, M. D. (2006). Assessment of airborne heavy metal pollution by aboveground plant parts. *Chemosphere,* 65, 177–182.

Saarela, K. E.; Harju, L.; Rajander, J.; Lill, J. O.; Heselius, S. J.; Lindroos, A. and Mattsson, K. (2005). Elemental analyses of pine bark and wood in an environmental study. *Science of The Total Environment*, 343, 231–241.

Salemaa, M.; Derome, J.; Helmisaari, H.; Nieminen, T. and Vanha-Majamaa, I. (2004). Element accumulation in boreal bryophytes, lichens and vascular plants exposed to heavy metal and sulfur deposition in Finland. *Science of The Total Environment*, 324 (1–3), 141–160.

Samecka-Cymerman, A.; Kolon, K. and Kempers, A. J. (2009). Short shoots of *Betula pendula* Roth. as bioindicators of urban environmental pollution in Wrocław (Poland). *Trees,* 23 (5), 923–929.

SEPA (Serbian Environmental Protection Agency) (2003-2010). Environmental reports for the republic of Serbia 2003–2010. The Ministry of Environment, Mining and Spatial Planning, Republic of Serbia, Belgrade (in Serbian). Available from: http://www.sepa.gov.rs/ index.php?id= 13andakcija= showDocsAll

Serbula, S. M. and Stankovic, V. D. (2002). Influence of an electrochemically generated gas phase on the pressure drop in a three–dimensional electrode and an inert bed. *Journal of Applied Electrochemistry*, 32, 389–394.

Šerbula, S. M.; Antonijević, M. M.; Milosević, N. M.; Milić, S. M. and Ilić A. A. (2010). Concentrations of particulate matter and arsenic in Bor (Serbia). *Journal of Hazardous Materials*, 181 (1–3), 43–51.

Šerbula, SM; Stevanović, J.; Trujić, V. Title: Arsenic, Heavy Metals and SO_2 Derived in a Mining-Metallurgical Production Process. In: Brar SK editor. Title: Hazardous Materials: Types, Risks and Control. New York: Nova Science Publishers US; 2011; 187–223.

Serbula, S. M.; Miljkovic, D. Dj.; Kovacevic, R. M. and Ilic, A. A. (2012a). Assessment of airborne heavy metal pollution using plant parts and topsoil. Ecotoxicology and Environmental Safety, 76 (1), 209–214.

Serbula, S. M.; Alagic, S. C; Ilic, A. A.; Kalinovic, T. S. and Strojic, J. V. Title: Particulate Matter Originated From Mining-Metallurgical Processes. In: Knudsen H, Rasmussen N. editors. Title: Particulate Matter: Sources, Emission Rates and Health Effects. New York: Nova Science Publishers US; 2012b; 9–116.

Serbula, S. M.; Kalinovic, T. S.; Kalinovic, J. V. and Ilic, A. A. (2013a). Exceedance of air quality standards resulting from pyro-metallurgical production of copper: a case study, Bor (Eastern Serbia). *Environmental Earth Sciences,* 68 (7), 1989-1998.

Serbula, S. M.; Kalinovic, T. S.; Ilic, A. A. and Kalinovic, J. V. and Steharnik, M. M. (2013b). Assessment of airborne heavy metal pollution using *Pinus spp*. and *Tilia spp*. *Aerosol and Air Quality Research*, 13 (2) 563–573.

Singh, R.; Singh, D. P.; Kumar, N.; Bhargava, S. K. and Barman, S. C. (2010). Accumulation and translocation of heavy metals in soil and plants from fly ash contaminated area. *Journal of Environmental Biology*, 31 (4), 421-430.

Steinnes, E.; Lukina, N.; Nikonov, V.; Aamlid, D. and Royset, O. (2000). A gradient study of 34 elements in the vicinity of a copper-nickel smelter in the Kola Peninsula. *Environmental Monitoring and Assessment*, 60 (1), 71–88.

Stojakovic, Dj.; Bugarski, B. and Rajic, N. (2012). A kinetic study of the release of vanillin encapsulated in Carnauba wax microcapsules. *Journal of Food Engineering,* 109, 640–642.

Sucharova J. and Suchara I. (2004). Distribution of 36 element deposition rates in a historic mining and smelting area as determined through fine-scale biomonitoring techniques. Part I: Relative and absolute current atmospheric deposition levels detected by moss analyses. *Water, Air, and Soil Pollution,* 153 (1), 205–228.

Sun, F. F.; Wen, da Z.; Kuang, Y. W.; Li, J. and Zhang, J. G. (2009). Concentrations of sulphur and heavy metals in needles and rooting soils of Masson pine (Pinus massoniana L.) trees growing along an urban-rural gradient in Guangzhou, China. *Environmental Monitoring and Assessment*, 154 (1–4), 263–274.

The Official Gazette of The Republic of Serbia (2010) Regulation on alterations and anexes of the Regulation on conditions for monitoring and the requirements for air quality improvement. No. 75/10

Tomašević, M.; Aničić, M.; Jovanović, Lj.; Perić-Grujić, A. and Ristić, M. (2011). Deciduous tree leaves in trace elements biomonitoring: A contribution to methodology. *Ecological Indicators*, 11 (6) 1689–1695.

Tretiach, M.; Pittao, E.; Crisafulli, P. and Adamo, P. (2011). Influence of exposure sites on trace element enrichment in moss-bags and characterization of particles deposited on the biomonitor surface. *Science of The Total Environment,* 409 (4), 822–830.

US EPA (U.S. Environmental Protection Agency) (2010). National Ambient Air Quality Standards (NAAQS) USA.

US EPA (1996). Method 3050B Acid Digestion of Sediments, Sludges and Siols. Available from: http://www.epa.gov/ osw/hazard/ testmethods/ sw846/pdfs/3050b.pdf

WHO (World health organization), (2000). Air Quality Guidelines for Europe. Copenhagen.

Wolterbeek, B. (2002). Biomonitoring of trace element air pollution: principles, possibilities and perspectives. *Environmental Pollution,* 120 (1), 11–21.

Yilmaz, S. and Zengin, M. (2004). Monitoring environmental pollution in Erzurum by chemical analysis of Scots pine (*Pinus sylvestris L.*) needles. *Environment International*, 29 (8), 1041–1047.

In: Air Quality
Editor: Arthur Hermans

ISBN: 978-1-62808-259-3

Chapter 4

ECOTOXICITY ASSESSMENT OF TRAFFIC-RELATED AIRBORNE POLLUTION

Nora Kováts*[*] *and András Gelencsér
University of Pannonia, Institute of Environmental Sciences,
Veszprém, Hungary

ABSTRACT

Traffic-related airborne pollution has been well characterised from (human) toxicological aspects, also, the role of plants in active/passive biomonitoring has been extensively investigated. On the contrary, much less information is available on the ecotoxicity of these emissions. Ecotoxicological tests (or bioassays) are controlled, reproducible tests in which ecological responses are determined quantitatively. These ecological responses are termed test end-points. Mortality as an ultimate (and easy-to-quantify) end-point is widely used, but other, sub-lethal endpoints exist such as impairment of growth/development, reproductive and/or biochemical dysfunctions, morphological abnormalities, or biochemical responses. Most authors agree that bioassays can be used in biomonitoring. Bioassays, however, can not only provide quantitative information on the toxic effect, but a clear concentration-response relationship can be established and concentration-response or stressor-response patterns can be analysed.

[*] University of Pannonia, Institute of Environmental Sciences, Egyetem Str. 10, H-8200 Veszprém, Hungary, Phone +36-88-624-317, Email: kovats@almos.uni-pannon.hu.

For ecotoxicity testing, the *Vibrio fischeri* bioluminescence inhibition test has been used almost exclusively. This chapter gives an overview of the basic principles of the bioassay, gives examples on its application for ecotoxicity assessment of airborne pollution, and also includes state-of-the art developments such as lux-based biosensors or the kinetic version of the test. Apart from this bioassay, other sporadically applied methods are described.

1. INTRODUCTION

Airborne particulate matter (PM) is usually classified by its aerodynamic diameter into coarse (≤10 µm, PM_{10}) and fine (≤ 2.5µm, $PM_{2.5}$) modes (Gualtieri et al., 2009). Diesel exhaust is one of the major sources of fine and ultra-fine (<100 nm in diameter) particulate matter ($PM_{2.5}$ and $PM_{0.1}$) in urban air. It was estimated that in large cities up to 50% of $PM_{2.5}$ mass concentrations may result from traffic-related emissions and resuspension of road dust (Chen et al., 2001; Schauer et al., 1996). Concentration of the respirable fraction (PM_{10}) is widely used as an air quality indicator (WHO, 2006; EC, 2008).

Human health impact of urban and/or traffic-related particulate matter has been widely studied either in epidemiological studies or in laboratory *in vivo*/*in vitro* tests. Much less information is available on the ecotoxic effects of these stressors on the non-human biota, ecosystems, etc.

2. BIOINDICATORS VS. ECOTOXICITY ASSESSMENT

The use of bioindicator/biomonitor organisms to assess the harmful effects of airborne pollution has been well documented. A short introduction is necessary to explain the distinction between ecotoxicological assays and bioindicator/biomonitoring studies.

Bioindicators can be divided into groups of effect or impact and accumulation indicators (Kovács, 1992; Franzle, 2006; Markert, 2007). Effect indicators show some detectable symptom, while accumulation indicators take up and accumulate elements or compounds in their monitor tissues (Wolterbeek, 2002).

Also, a distinction between active and passive bioindication is necessary. Active bioindication implies the intentional exposure of selected organisms in a standardised form to environmental stress, whilst passive bioindication is actually a field survey where stress reactions are evaluated in naturally occurring elements of the ecosystem.

It is generally agreed that bioindicators are organisms which can give qualitative information on the status of their environment, while biomonitors are organisms which are used for the quantitative determination of environmental impacts (Conti and Cecchetti, 2001; Markert et al., 2003).

Most often, biomonitoring implies that concentration of selected elements is measured from plant tissues. Lichens were the first organisms used in the assessment of air pollution and are still used in modern studies (e.g. Carreras and Pignata, 2002; Marques et al., 2004; Ayrault et al., 2007; Bergamaschi et al., 2007; Brunialti and Frati, 2007; Branquinho et al., 2008; Godinho et al., 2009).

Another important group is mosses: both lichens and mosses are very sensitive to air quality, as they do not have epidermis or cuticula. In many cases, these two taxa are used in parallel (Tretiach et al., 2007) Apart from individual studies, complete reviews are available on lichen and/or moss based biomonitoring (Conti and Cecchetti, 2001; Szczepaniak and Biziuk, 2003).

Some higher plants have also been used, such as *Tillandsia* species (Bromeliaceae) (Gonzalo et al., 2009), Italian rye grass (*Lolium multiflorum* Lam., Poaceae) (Klumpp et al., 2009). Tree bark has also proven to be a reliable indicator (e.g. Rusu et al., 2006; Baptista et al., 2008).

However, it should be stressed that analytical determination of the elements (metals) taken up by the plant will not give any indication on the ecological effect of the pollutant in question.

Ecotoxicological tests (or bioassays) are controlled, reproducible tests where ecological responses are determined quantitatively. These ecological responses are termed test end-points (Suter, 1993). Mortality as an ultimate (and easy-to-quantify) end-point is widely used, but other, sub-lethal endpoints exist such as impairment of growth/development, reproductive and/or biochemical dysfunctions, morphological abnormalities, or biochemical responses. Most authors agree that bioassays can be used in biomonitoring (Franzle, 2006; Markert, 2007). Bioassays, however, can not only provide quantitative information on the (eco)toxic effect, but ecotoxicological tests are aiming at determining the relationship between the level of environmental exposure (the quantity to which the organism is exposed to) and the nature and level of the ecological effects (Walker et al., 2006). Ecotoxicity of the sample is most commonly expressed in the form of EC_x value, that is, the (calculated) effective concentration which causes x % of ecological effect. Most often,

EC_{50} (concentration causing 50% of effect) is used. Apart from expressing ecotoxicity of the sample as a single metric, clear concentration-response relationship can be given and concentration-response or stressor-response patterns (USEPA, 2000) can be analysed.

3. Ecotoxicological Assays

Contrary to toxicology, ecotoxicology is concerned with effects on whole ecosystems (Moriarty, 1983). Important ecosystem structural and/or functional elements are represented in laboratory studies, using surrogate species: for example, trophic structure of an aquatic ecosystem is composed of the following levels: primary producers (represented by algae such as the standard *Scenedesmus capricornutus* test), primary consumers (surrogate species are Daphnia) and secondary consumers (represented by fishes).

Airborne pollution raises, however, several methodological constraints. Considering the nature of the material to be sampled, sampling itself is relatively difficult as most ecotoxicological assays require a relatively large sample quantity. Airborne emissions are normally collected on a filter, which limits the quantity of the sample and also, the range of available bioassays. The fact is that normally low amount of PM can be collected in many cases limits the evaluation of PM toxicity as well as argued by DeVizcaya-Ruiz et al. (2006).

As such, for ecotoxicity testing bacterial bioassays have been used, most frequently the *Vibrio fischeri* bioluminescence inhibition bioassay. Originally, *V. fischeri* can be found in small amounts in the ocean and in large amount in isolated areas such as the light organs of a squid, *Euprymna scolopes* with which it has a symbiotic relationship. When the squid is young, it draws in free-living bacteria from the ocean into its light organ. Here they are provided with all of the nutrients that they need to survive. Light emittance is activated only within the squid, as in the ocean cell density is app. 10^2 cells/ml, and this low concentration of cells is not enough to cause the luminescence genes to be activated.

Cell density-dependent control of gene expression of lux genes is activated by autoinduction that involves the coupling of a transcriptional activator protein with a signal molecule (autoinducer) that is released by the bacteria into its surrounding environment (this ‘communication’ is called quorum-sensing[1]). When in the light organ of a squid, the cell concentration is about 10^{10} cells/ml, and the autoinducer causes the bacteria to emit light. The squid is even capable of controlling light emittance: during the day, it keeps the bacteria at lower concentrations by expelling some of them into the ocean during regular intervals. At night, as the squid is night-feeder, the bacteria are 'allowed in'.

Under laboratory conditions, naturally, test organisms are having enough density to induce light emittance. The end-point of the test is the reduction of light output which is proportional to the strength of the toxin (Fig. 1.).

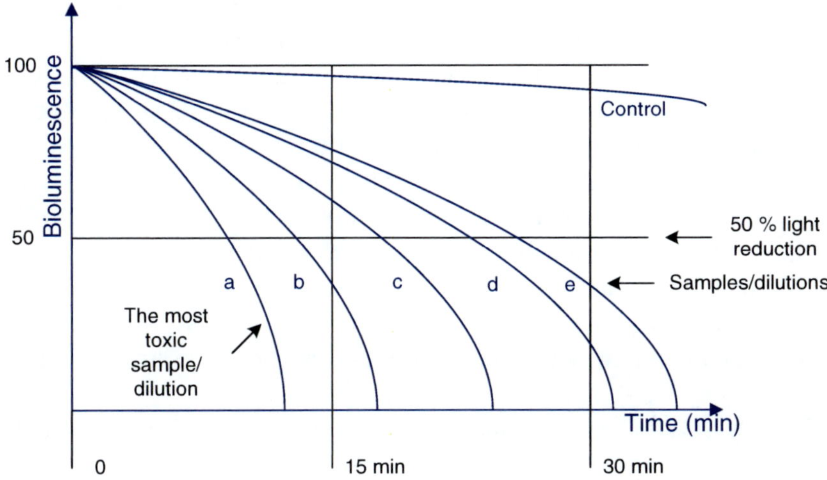

Figure 1. Luminescence inhibition is roughly proportional to the concentration of the toxic compound.

The earliest commercial application of this biochemical reaction dates back to 1981, when Beckman Instruments introduced its system which is presently marketed under the name 'Microtox' (Microbics Corporation, Carlsbad, CA, later on AZUR Environmental) (Bulich and Isenberg, 1981). Other commercial systems exist such as ToxAlert (Merck), LUMIStox (Hach Lange), BioTox (ABOATOX). There are several standardized protocols,

[1] The regulation of density-dependent behaviour by means of quorum sensing is widespread in bacteria. It can be used for example by opportunistic/infectious bacteria such as *Pseudomonas aeruginosa* to overcome the host's immune system: bacteria grow to a certain concentration without any warning sign and they become aggressive only when reaching a critical mass.

proposed first by Bulich (1979) and applied by national (e.g. AFNOR, 1991) or international organizations (ISO, 1998). Results obtained using different systems show little variance (Jennings et al, 2003).

As bioluminescence is directly linked to respiratory activity, it gives a good indication on the metabolic activity of the organism and was found to show good correlation with several *in vivo* toxicity tests. Fort (1992) found that for the purpose of monitoring toxins in pharmaceuticals, Microtox EC_x values could be correlated with the LD (lethal dose) values from assays using mouse survival. Lebsack et al. (1981) compared the Microtox test to bioassays using Rainbow trout and Fathead minnows (24–96 hour exposure). These bioassays were evaluated with fossil-fuel process waters containing ammonia and phenolic constituents. There was a substantial correlation in toxic effects between the three tests. Burton et al. (1986) conducted a comparison of bacterial bioluminescence tests, tissue culture bioassays, and an *in vivo* test for evaluating acute toxicity of biomaterials (plastics). Tissue cultures were used to assess the cytologic changes that occur in cells exposed to leachable constituents. In vivo tests involve implanting a material directly into a living rabbit. The tissue surrounding the implant is then evaluated after several days for hemorrhage, film formation, or encapsulation. In order to apply the Microtox test the plastic samples were extracted with the Microtox culture medium. In this application the Microtox test was more sensitive than either *in vivo* test.

The *V. fischeri* bioassay has been shown to exert proper sensitivity to contaminants expected to occur in airborne particles, most important group being PAHs and metals. Sensitivity of the *V. fischeri* assay to PAHs has been established (El-Alawi et. al., 2002; Hirmann et al., 2007), even EC_{50}s of individual PAHs are available (Eom et al., 2007). El-Alawi et al. (2001) assessed the photoinduced toxicity of PAHs and found that simulated solar radiation enhanced the toxic potential of tested PAHs. Pattern of metal toxicity on *V. fischeri* is discussed in details by Fulladosa et al., 2005.

3.1. Bulk Samples

The conventional protocol uses aqueous samples (in compliance with ISO 11348). In case of solid samples (most often sediment or soil), an elutriate should be prepared. Although several authors successfully applied elutriates for characterising solid media contamination (e.g. Salizzato et al., 1998; Olajire et al., 2005; Ocampo-Duque et al., 2008; Giltrap et al., 2009), there are some methodological constraints. The biggest limitation of using extracts, however, is that the toxic effect depends mostly on particle-bound compounds (Harkey and Young, 2000).

For solid-phase (sediment and soil) testing, Brouwer et al. (1990) and Tung et al. (1990) separately proposed a protocol which uses direct contact between bacteria and solid phase. This protocol was modified by Microbics Corporation (1992) and was proposed by Azur Environmental (1998) as a standard for testing solid samples. In the Microtox® Solid Phase Basic Test the bacteria are mixed with different amounts of the test sediment. After the exposure period, the solid particles are removed by filtration and in fact, toxicity of the water fraction is estimated. One major constraint, as reported by some authors is that a given portion of the bacteria might be lost due to their adhesion onto the sample matrix, resulting in a lower light emission reading (Ringwood et al., 1997). Also, turbidity and colour of sediments might cause a significant decrease of light output due to physical effects, creating the potential for false-positives (Campisi et al., 2005).

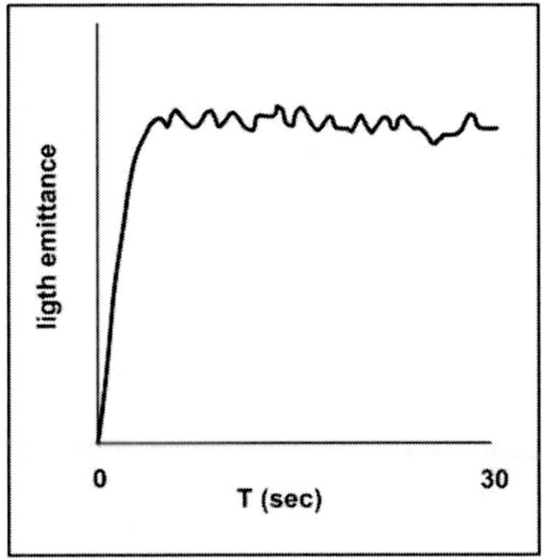

Figure 2. Typical kinetic behaviour of light emittance in the control during 30 s of exposure.

Lappalainen et al. (1999, 2001) presented a novel protocol for testing the toxicity of solid and/or coloured samples. Here luminescence intensity is evaluated in a kinetic mode. As the bacterial suspension is injected to the sample, the luminous intensity increases, to a peak (maximum) within 30 s (that is why the system is called Flash). The light emission is measured and recorded continuously during this 30 s, from the moment of dispersing of the bacterial suspension in the sample until the maximum value has been reached. Typical kinetic curves already show the 'behavioural pattern' of different samples (Fig 2 -5.).

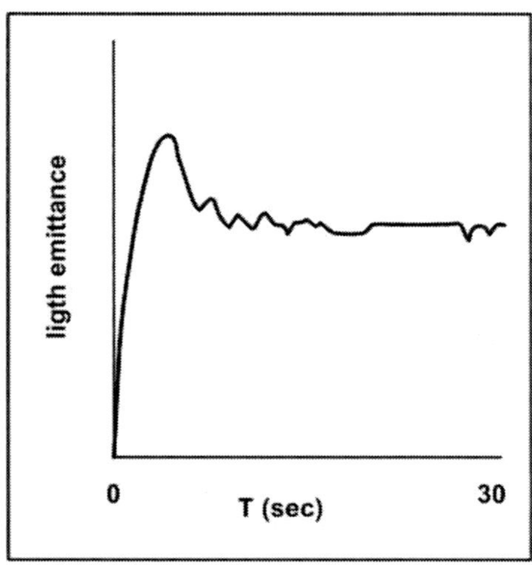

Figure 3. Typical kinetic behaviour of light emittance in a toxic sample during 30 s of exposure.

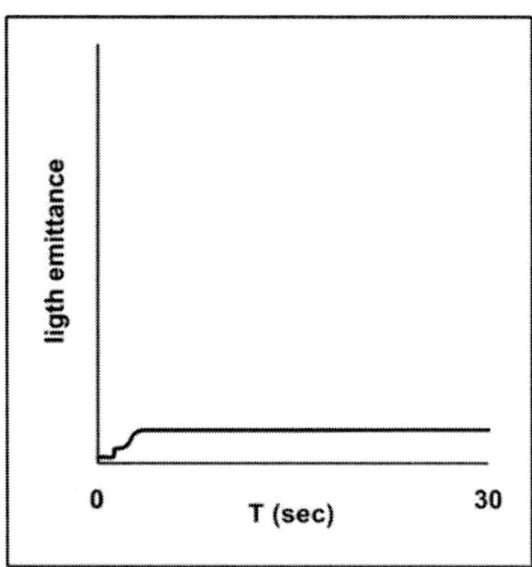

Figure 4. Typical kinetic behaviour of light emittance in a nontoxic, turbid and/or coloured sample during 30 s of exposure.

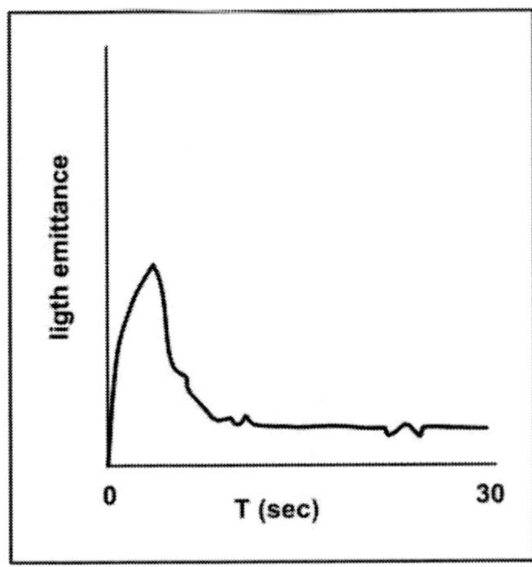

Figure 5. Typical kinetic behaviour of light emittance in a toxic, turbid and/or coloured sample during 30 s of exposure.

This kinetic recording of luminescence during the first 30s (which is continuously monitored by the system) can already give some indication on the behaviour of the sample (Mortimer et al., 2008).

Inhibition can be calculated comparing the initial and final readings, independently from the control. As such, inhibition calculation also differs considering the conventional protocol and the kinetic assay (Table 1.)

Table 1. Bioluminenscence inhibition calculation in the conventional protocol and in the kinetic assay

Conventional protocol	Kinetic assay
Firstly the f_{kt} correction factor is calculated from the measured luminescence: $f_{kt} = l_{kt} / l_0$ Where f_{kt} the correction factor for the contact time l_{kt} luminescence intensity in the control sample measured in RLU (relative luminescence units), after the contact time l_0 luminescence intensity of the control test suspension. Using the correction factor, than the corrected value of for every test sample cuvettes are calculated (Equation [2]).	$KF = IC_{15}/IC_0$ INH% = 100 – 100 x (IT_{15} / KF x IT_0) Where: KF = correction factor. IC_{15} = luminescence intensity of control after contact time (15 min) in RLU. IC_0 = initial luminescence intensity of control sample in RLU. IT_{15} = luminescence intensity

Table 1. (Continued)

Conventional protocol	Kinetic assay
$l_{ct} = l_0 \times f_{kt}$ [2] where f_{kt} mean of f_{kt} of the two control samples l_0 luminescence intensity of the control test suspension l_{ct} corrected value of for test sample cuvettes immediately before the addition of the test sample Then the inhibitory effect H_t of the test sample is calculated (Equation [3]). $H_t = [(l_{ct} - l_{Tt}/l_{ct})] \times 100$ [3] where H_t the inhibitory effect of the test sample after the contact time, in% l_{ct} corrected value of for test sample cuvettes immediately before the addition of the test sample l_{Tt} luminescence intensity of the test sample after the contact time, in RLU.	of test sample after contact time (15 min) in RLU. IT_0 = initial luminescence intensity of the test sample in RLU.

The kinetic test protocol has been standardised, the ISO standard (ISO 21338:2010: Water quality - Kinetic determination of the inhibitory effects of sediment, other solids and coloured samples on the light emission of *Vibrio fischeri* /kinetic luminescent bacteria test/) was issued in 2010.

3.1.1. Assessing Elutriate Ecotoxicity Using Microtox

Despite all the drawbacks, the most often applied sample preparation method is taking an elutriate or extract which is further used as an aqueous solution for the *V. fischeri* bioassay. Triolo et al. (2008) used the Microtox® system for the ecotoxicological evaluation of airborne pollutants. The main aim of their study was to assess the environmental health effect of industrial emission on agriculture in the vicinity of Milazzo (Italy). Apart from ecotoxicological assays, in the frame of a comprehensive monitoring, as exposure indicators, sulphur dioxide and ozone levels in air were monitored, and trace element concentrations in crops and soil were analysed. As impact indicators, metabolic profiles of soil microbial communities were taken. For the Microtox assay, organic contaminants were collected on Semipermeable

Membrane Devices (SPMDs). SPMDs were rinsed and solubilised with acetone: DMSO mixture. The acetone was evaporated by nitrogen, and the recovered sample was used (Table 2). In fact, no clear stressor-response relationship could be found between primary pollutants and ecotoxicity, the authors hypothesize that other pollution sources such as pesticide use can influence the pattern of ecotoxicity. Roig et al. (2013) also applied Microtox assay to assess the ecotoxicity of airborne pollutants, on samples collected in different parts of Catalonia (Spain). Assays were performed on aqueous extract of aerosol filters after mild acid microwave digestion (15 mL 0.1 M HNO_3 at 2% NaCl). After the chemical analysis of the samples, most pollutants showed significant correlation with toxicity values (such pollutants were metals and polychlorinated dibenzo-p-dioxins and dibenzofurans /PCDD/Fs/). In their study, toxicity showed a spatio-temporal pattern, with urban samples being more ecotoxic than rural samples and winter samples being more ecotoxic than summer samples. Vouitsis et al. (2009) estimated the ecotoxicity of particulate matter (PM) emitted from light-duty vehicles using the Microtox system. In order to prepare liquid phase (*V. fischeri* are marine organisms), filters collecting exhaust samples were Soxhlet extracted in 300 mL dichloromethane (DCM) for 24 h followed by rotary evaporation, to reduce the volume to 1 mL. The concentrated extract was then exchanged with 1 mL dimethylsulphoxide (DMSO) and was diluted to a final volume of 10 mL with deionised water. In their study, PM emission of single vehicles (one diesel and one gasoline car) was characterised, comparing different driving cycles and in case of the diesel car, the vehicle was first tested in the original exhaust aftertreatment (Diesel OEA) configuration, then after replacing the underfloor catalyst with a silicon-carbide catalysed diesel particle filter (CDPF). Ecotoxicity of the emission decreased in the following order: gasoline car > CDPF > Diesel OEA. No significant difference was found, however, for different driving cycles.

Table 2. Sample preparation for Microtox assays

Triolo et al. (2008)	Roig et al. (2013)	Lin and Chao (2002) Vouitsis et al. (2009)
Samples collected on standard Semipermeable Membrane Devices (SPMDs) SPMDs are rinsed to remove surface fouled residues then solubilised with 20 mlLof an acetone:dymethylsulphoxide	For sampling, TE-1000PUF and TE-6070DV (Tisch Environmental, Cleves, OH) high-volume active sampling device used	PM collected on 47 mm PTFE-coated fibre filters (Pallflex TX40H120-WW) Filters collecting exhaust gas samples are Soxhlet extracted in 300 mL

Table 2. (Continued)

Triolo et al. (2008)	Roig et al. (2013)	Lin and Chao (2002) Vouitsis et al. (2009)
(DMSO) mixture (1:1v/v) for 24h. The acetone is evaporated by nitrogen and the DMSO extract is used for toxicity test	Aqueous extraction of aerosol filters is performed by mild acid microwave digestion (15mL 0.1MHNO3 at 2% NaCl) After microwave digestion, the content is filtered with Whatman®-42 filters, the resulting extract neutralized to pH7.0±0.2 with0.5M NaOH	dichloromethane (DCM) for 24 h followed by rotary evaporation, to reduce the volume to 1 mL The concentrated extract is then exchanged with 1 mL dimethylsulfoxide (DMSO) and is diluted to a final volume of 10 mL with deionised water

Lin and Chao (2002) used a similar sample preparation method to assess the influence of methanol-containing additive on acute toxicity of diesel exhaust emissions. In their study, hot start and steady-state modes of a heavy duty vehicle were compared, assessing the acute toxicity of the vapour-phase (XOC) and particle associated soluble organic fraction (SOF) samples. They found that in general, strong correlation could be found between XOC-associated toxicity and total hydrocarbon concentrations, actually, toxicity of vapour-phase (XOC) samples was higher than that of SOF samples in all cycle tests. Microtox assays also showed that addition of moderate amount of methanol-containing additive (5 or 8%) did lower the SOF-associated toxicity but higher portion of the additive (10-15%) increased SOF toxicity in comparison to the base fuel.

3.1.2. Direct Contact Test

The kinetic Flash test was applied by Kováts et al. (2012), combined with a specific novel sample preparation method. Filter sample spots of 25 mm in diameter are cut with a special puncher then measured gravimetrically with a Sartorius microbalance (10 µg sensitivity). Filter spots are ground in an agate mortar then transferred into pre-cleaned 4 ml vials with a PTFE-coated spatula. Then suspensions are prepared with 2 mL high-purity (MiiliQ) water. In the suspension, bacteria are in direct contact with the toxic particles. In order to avoid quality assurance problems, filters were previously checked and proved inert, therefore all toxicity readings can be attributed to the sample collected on the filter. EC_{50}s are basically calculated as % of the suspension, but as mass of aerosol collected on the filter is measured, EC_{50}s can be calculated as mg aerosol (Turóczi et al, 2012).

This protocol was first applied to assess the potential ecotoxicity of a red mud spill which occurred in 2010 in Hungary (Gelencsér et al., 2011).

Figure 6. Sampling device.

The team had previously developed a sampling device for collecting the respirable fraction of resuspended road dust (PM_{10}) (Turóczi *et al.*, 2011). The instrument uses a leaf blower to mobilize dust from paved surfaces, simulating very windy conditions (wind speeds 100 km h-1). The key unit in the apparatus is a PARTISOL-FRM model 2000 sampler that is operated at a flow rate of 16.7 L min-1 and contains a cyclone separator which collects the $PM_{(10\text{-}1)}$ fraction (particles with aerodynamically equivalent diameters between 10 μm and 1 μm) in bulk in a glass holder. The sampler is mounted on a mobile platform and powered with a portable electrical power generator (Fig 6.).

Based on this method, Turóczi et al. (2012) performed a comparative assessment of ecotoxicity of urban aerosol, with special regard to seasonal differences. They found that ecotoxicity of winter and summer aerosol samples showed a striking difference. The much higher ecotoxicity of winter samples might be explained by the high contribution of vehicular emissions from a rather aged vehicle fleet. In Budapest, the average ages of LDVs and HDVs were 10 and 12 years in 2010, respectively. The percentage of buses belonging to different emission standard classes was as follows, with reference to 2010' conditions (Varga et al., 2010): Euro0: 35%; Euro1: 9%; Euro2: 17%; Euro3: 23%; Euro4: 12% and Euro5: 4%. Another important factor could be the similarly high share of wood burning emissions in winter which are also ecotoxic (Gelencsér et al., 2007). Other factors which might possibly enhance winter ecotoxicity are lower rates of photooxidation, and lower degree of atmospheric mixing due to the low mixing height and frequent inversions. In addition, condensation of semi-volatile compounds onto the particles is favoured at low temperatures in winter.

A thorough test series was conducted to verify the applicability of the direct contact test protocol for assessing the environmental load of individual vehicles (Turóczi et al., 2012, Ács et al., unpublished data, Kováts et al., 2013). In the first study (Turóczi et al., 2012, Kováts et al., 2013), ecotoxicity of the emission of six in-service public transport buses was assessed. The age of selected buses ranged from 3 to 24 years and also, different emission standard classes were included, from Euro0 to Euro4. Particulate samples from the exhausts were collected with a KÁLMÁN $PM_{2.5}$ sampler at a flow rate of 32 m^3 h^{-1} for 10 minutes at idling (600 rpm) and for 3 minutes with slightly open throttle (1500 rpm) in a closed garage. EC_{50}s were calculated as % of the suspension then recalculated as µg/mL (µg of aerosol/mL suspension).

At idling, EC_{50}s of the particulate emission of the vehicles ranged from 0.96 µg/ml belonging to Euro1 to non-toxic (Euro4). Surprisingly, EC_{50} values of Euro0 and Euro1 emissions, as well as that of one Euro3 bus fell very close to each other. It is interesting to note that in parallel, *Artemia salina* (crustacean) bioassays (typical mortality tests) were conducted on the same samples. The results of the test showed similar trend as the Flash assay, however, sensitivity of this test organism was several magnitudes lower (unpublished data). Though no other comparison exists between the sensitivity of *V. fischeri* and other (standard) test organisms for total aerosol samples, Eom et al. (2007) compared the sensitivity of *V. fischeri* to individual PAHs with two other standard, invertebrate tests (*Daphnia* and *Ceriodaphnia*) and found that the sensitivity ranking of these tests based on EC_{50} values was, in decreasing order: *Ceriodaphnia* > Microtox (*V. fischeri*) > *Daphnia*.

In another similar study, aerosol samples from the exhausts of 9 diesel-powered passenger cars were assessed for ecotoxicity (Ács et al., unpublished data). The same sampling method was applied as described for the previous study. Other parameters were also measured: total carbon (TC), with a Zellweger Astro TOC 2100 total carbon analyzer and total PAH content. The $PM_{2.5}$ mass concentrations of diesel engine exhaust samples were determined indirectly by total carbon (TC) mass measurements, using the TC/$PM_{2.5}$ ratio = 0.92 for passenger cars (Graham et al., 2005). EC_{50} values for each sample were calculated than converted into Toxic Units (TU). The Toxic Unit approach was originally developed to test the response addition model for chemical mixtures (Pape-Lindstrom and Lydy, 1997), but is has also been used for presenting the toxicity of single samples (e.g. Triolo et al., 2008). TU is a dimensionless value given as 100/ EC_{50}. Significant positive correlation was found between ecotoxicity (TU values) and TC (R=0.912; P= 0.001) as well as between TU and PAH concentrations (R=0.956; P= 0.001). Seemingly, ecotoxic effect of the diesel exhaust depended directly on the quantity of carbonaceous particles produced, which might act as carriers for toxic compounds covering these particulates. Also, emission of individual vehicles could be characterized.

The conventional bioluminescence inhibition bioassay in which an extract is used might not reflect a realistic environmental exposure route as by the use of organic solvents under extreme conditions the toxicity of the aerosol might be overestimated as many components are mobilized which are normally not bioavailable. On the other hand, in systems using the conventional protocol false readings might occur if the sample is turbid and/or coloured, due to the Tyndall scattering.

Kováts et al. (2011) compared the ecotoxicity or urban aerosol samples using the two versions of the bioluminescence inhibition bioassay: ToxAlert®100 (ISO 11348-3:2007 protocol) and Flash (ISO 21338:2010 protocol). For the ToxAlert 100 bioassay, extracts were prepared as described by Vouitsis et al. (2009). The Flash system, as mentioned already, assesses the ecotoxicity of a suspension where bacteria are in direct contact with the previously grounded sample filter. Sensitivity of the two systems was very different: when winter samples were assessed, EC_{50} values fell rather close to each other, but in case of summer samples, ToxAlert showed detectable ecotoxicity while according to Flash readings, these samples were not ecotoxic at all.

3.1.3. Expression of Ecotoxicity

Ecotoxicity is generally expressed as EC_{20} or EC_{50}, that is, the concentration which causes 20 or 50 % of ecological effect (e.g., in case of the *V. fischeri* bioluminescence inhibition assay, the effect is the inhibition). Depending on the sampling method or the sample preparation procedure, these values may refer to different environmental media. For example, Lin and Chao (2002) calculated EC_{50}s as the extract concentration causing 50% of inhibition per $0.1m^3$ of raw exhaust sampled.

Voutsis et al. (2009) derived EC_{50}s from the diluted concentrations following extraction procedure. Toxicity was given as relative toxicity (TR), expressed as a ratio over the diesel OEA New European Driving Cycle (NEDC) (Andre, 2004).

Roig et al. (2013) calculated EC_{50}s from the aqueous extracts, but finally toxicity was expressed as a Toxicity Index (TI) calculated by comparing the EC50 of each sample with the value estimated for a blank filter (Roig et al., 2011).

Table 3. Expression of ecotoxicity in the Microtox assays

Triolo et al. (2006)	Roig et al. (2013)	Vouitsis et al. (2009)	Lin and Chao (2002)
EC_{50} expressed as the concentration of the DMSO extract causing 50% inhibition Toxicity finally expressed as TU ($100/EC_{50}$)	EC_{50} expressed as the concentration of the aqueous extract causing 50% inhibition Toxicity finally expressed as a Toxicity Index (TI) calculated by comparing the EC_{50} of each sample with the value estimated for a blank filter	EC_{50} expressed as the concentration of the DMSO solution causing 50% inhibition Toxicity is finally given as relative toxicity (TR), expressed as a ratio over the diesel OEA New European Driving Cycle (NEDC)	EC_{50} expressed as the concentration of the DMSO solution causing 50% inhibition EC_{50}s are finally given as per $0.1m^3$ of raw exhaust sampled

(Eco)toxicity can also be assessed in an indirect way, based on toxicity equivalency factors (TEFs) scales available in the literature. Lim et al. (2005) for example used benzo(a)pyrene equivalent (BAPeq) factors when comparing the influence of fuel composition and engine operating conditions on toxic effect of heavy-duty diesel buses.

3.2. Ecotoxicity Testing of the Gaseous Phase

Gil et al. (2000, 2002) developed a whole-cell biosensor for detecting ecotoxicity of gases. The system uses a recombinant bioluminescent *Escherichia coli* incorporated into agar gel which harbors a lac:luxCDABE fusion. The immobilized cell matrix maintains the acitivity of test organisms which are in direct contact with the toxic gas flowing through the system. Similarly to the original *Vibrio fischeri* test, toxicity is evaluated through the reduction of bioluminescence. (Fig. 7.)

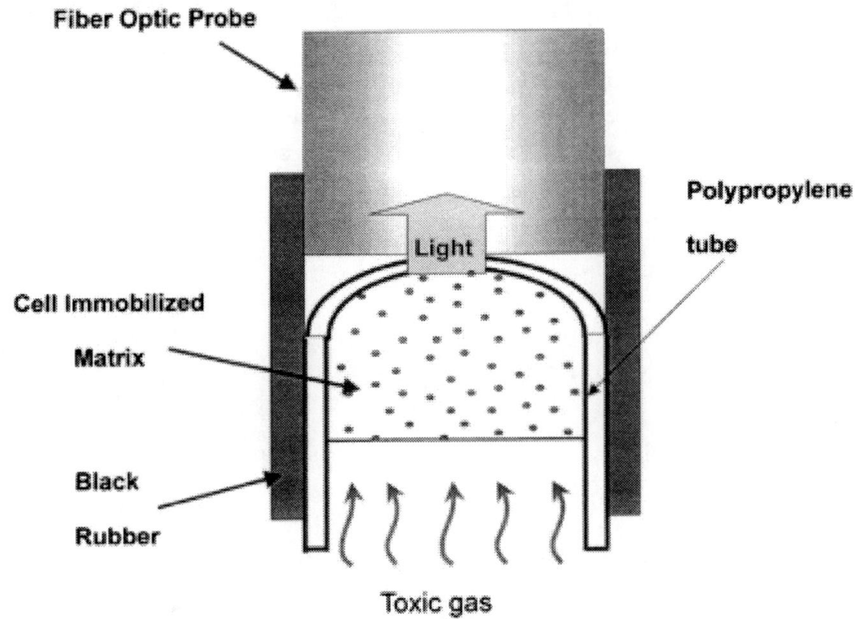

Figure 7. Schematic diagram of the immobilized cell matrix connected with a fiber optic probe (After Gil et al., 2000).

The system was first tested on benzene and a clear dose-response pattern was determined (Gil et al., 2000). Sensitivity to other BTEX gases such as toluene, ethylbenzene, and xylene) was proved (Gil et al., 2002). This biosensor could be produced in a compact size, functioning as a portable device.

Komori et al. (2009) developed a rapid gas toxicity evaluation system where *V. fischeri* are trapped using a polyion complex membrane in order to allow semi direct contact between the bacteria and toxic gases. The system gives EC_{50}s per ppm of model gases measured.

Sensitivity of the system was assessed on gas samples, such as benzene, trichloroethylene (TCE), acetone, NO_2, SO_2 and CO, as well as on emissions (diesel engine exhaust, gasoline engine exhaust, charcoal burner exhaust and cigarette smoke.) The authors also compared the sensitivity of this system to that of conventional animal tests and found that their gas sensor had a sensitivity 1–3 orders of magnitude higher than that of the animal tests. Of tested gases, volatile organic compounds most possibly cause damage to test bacteria based on the destruction of a phospholipid membrane due to permeation and/or accumulation (Schultz et al., 2003), while NO_2 andSO_2 gases are known to dissolve in aqueous media and then converted to nitric/nitrous and sulphuric acids, thus decreasing pH in the membrane which also has an inhibitory effect on the enzymatic activity of the bacteria. The main benefits of the system, in addition to its sensitivity, are the short exposure time (15 min, which is normal for *V. fischeri* bioassays) and also, a portable device can be based on this method.

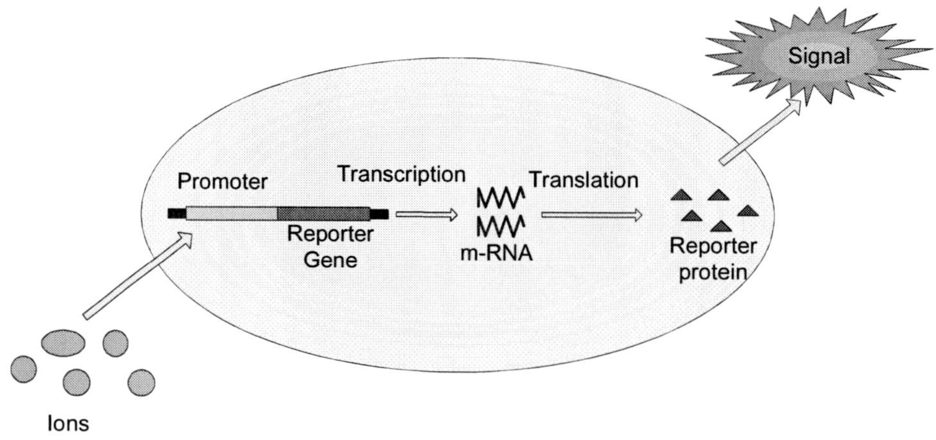

Figure 8. Sketch of a bioreporter organism. A specific analyte triggers the transcription of the promoter/reporter gene complex into messenger RNA which is then translated into a reporter protein. That protein induces the signal.

Biosensors producing bioluminescence inhibition (like the conventional *V. fischeri* protocols) are called 'light off' assays, however, there are biosensors which produce light signal: they are termed 'light on' assays (Kim et al., 2003). In 'light on' systems, the *lux* genes synthesize luciferase through the induction of a promoter triggered by chemicals, resulting in an increase in luminescence emission. These promoters are specific, responding to specific chemicals or specific stressors (Fig 8.).

Eltzov et al. (2011) used two different *Escherichia coli* strains in a fiber optic based biosensor for air toxicity monitoring. The DPD2794 strain harbours the recA promoter which activates DNA repair systems due to DNA damage (Vollmer et al., 1997; Elsemore, 1998; Davidov et al., 2000). The TV1061 strain harbours the heat-shock grpE promoter which is sensitive to cytotoxic substances (Arsene et al., 2000). In both strains, the analyte triggers the activation of the reporter luciferase genes and a detectable light signal is produced.

Valdman and Gutz (2008) also developed a biosensor for naphthalene in air, using *Pseudomonas fluorescens* HK44 strain. This strain harbours the bioluminescent reporter plasmid pUTK21 that contains a nahG-luxCDABE fusion in a salicylate inducible operon (D'Souza, 2001). The use of this strain has been well documented in soil bioremediation (Sayler et al.,1999; Ripp et al., 2000). This strain also has a low detection limit for naphtalane in air (Valdman et al., 2004). Its environmental relevance is high, as in large cities the highest concentration might reach 170 μg/m3 (Preuss et al., 2003). Also, recombinant bioluminescent bacteria such as the *E. coli*, strainRFM443, harbouring the pLITE2 plasmid (lac::luxCDABE), showed appropriate sensitivity against phenanthrene (Gu and Chang, 2001; Chang et al., 2004) and other PAHs such as naphthalene, anthracene, pyrene and benzo(a)pyrene (Lee et al., 2003).

4. Biochemical Markers

Biomarkers are measurements in body fluids, cells or tissues indicating biochemical or cellular modifications due either to the presence and magnitude of toxicants, or of recipient response, giving an early warning signal of the potential toxic effect (NRC, 1987). Biomarkers are classified into several groups, the most sensitive group is biotransformation enzymes. They are enzymes being involved in xenobiotic biotransformation and their activity can either be induced or inhibited upon exposure to xenobiotics.

The inhibition of acetylcholinesterase (AChE) activity, changes in lactate dehydrogenase (LDH) activity and the induction of glutathione S-transferases (GSTs) have been commonly used to assess the effects of different environmental stressors. AChE inhibition was first used in the effect assessment of organophosphate and carbamate pesticides (Peakall, 1992). GSTs are a family of enzymes which have a central role in the detoxification of oxidative stress products (George, 1994) and in the prevention of lipid peroxidation, via controlling the conjugation of glutathione (GSH) with certain xenobiotic compounds, increasing their water solubility and facilitating their excretion (Schlenk et al., 2008). The main function of LDH is to catalyse the reversible conversion of pyruvate to lactate (Vassault, 1983), playing a fundamental role in anaerobic pathway of energy production within the cell.

The relationship between biomarkers and PAHs have been most often studied in different media, such as oil-contaminated sediments or soils (e.g. Francioni et al., 2007; Brinkmann et al., 2013), oil spills (e.g. Marigómez et al., 2006), spiked sediments or in some cases, individual PAHs were used to elucidate ecological effect. Enzyme activity can be assessed by measuring protein levels and by measuring mRNA expressions of oxidative stress genes. Nahrgang et al. (2009) found when assessing the ecotoxic effect of benzo(a)pyrene that both the protein levels and mRNAexpression showed dose-response pattern, but protein levels gave weaker response.

However extensively biomarker responses to PAH exposure have been demonstrated, unfortunately there are no references about what biochemical/genetic responses aerosol samples would elucidate. It is important to note, that as aerosols generated by road transport are very complex systems, containing a set of toxic compounds, the aggregate ecotoxic effect has to be assessed, taking into consideration possible additive/synergistic/antagonistic effects amongst these contaminants.

5. Tire Debris

Apart from exhaust emission, tire and break wear are also an important source of respirable particulate matter in cities, contributing to app. 3–7% of PM emission originating from road transport (Gualtieri et al., 2005a). Wik and Dave (2009) give an overview about measured, estimated, and reported maximum concentrations of tire wear particles in various environmental compartments. Soil seems to be the primary compartment affected, aquatic ecosystems might be affected via soil contamination or via road runoff.

Although under 'normal' driving conditions less than 5% of tire-wear is emitted as PM_{10} (Wik and Dave, 2009), its potential toxicity cannot be neglected. Tire-wear contains potentially toxic components, such as PAHs (Takada et al.,1991) and metals, especially zinc (Councell et al., 2004). Water extracts of tires have been tested on organisms of different taxonomical orders, such as crustaceans (Day et al., 1993; Nelson et al., 1994; Gualtieri et al., 2005b; Wik and Dave, 2005, 2006; Marwood et al., 2011), amphibians (Gualtieri et al., 2005a; Mantecca et al., 2007); fish (Day et al., 1993; Nelson et al., 1994; Hartwell et al., 2000; Marwood et al., 2011), as well as algae (Gualtieri et al., 2005b; Marwood et al., 2011) and bacteria (Day et al., 1993; Hartwell et al., 2000).

In these studies either whole tyre or tyre pieces were extracted (Wik and Dave /2009/ give an overview about leaching procedures used in different ecotoxicological studies). Although no ecotoxicological testing has been conducted on the breathable fraction of tyre wear particles, the fact that in the abovementioned studies leachetes using tyre pieces showed higher toxicites than whole tyre extract might draw the attention to the potential risk of small particles.

CONCLUSION

Ecotoxicology, as defined previously, intends to assess the ecological effect of different contaminants, most possibly giving a quantitative estimation. Final recipients of environmental stressors are ecosystems, which are made of elements (populations), showing different sensitivity to the specific contaminant. Naturally, one and only one test organism – such as the *V. fischeri* bioluminescence inhibition test – cannot provide relevant information on the expected reaction of a whole ecosystem to airborne contaminants. The *V. fischeri* assay should be used for screening purposes or to compare the ecotoxicity of different samples such as emissions within a group of vehicles or emissions generated using different driving cycles, or fuels. Nevertheless, taking into consideration methodological constraints such as low sample quantity, sample preparation difficulties (all addressed to previously), most probably no such battery of tests can be composed which represents different organizational and/or functional and/or structural elements of the ecosystem.

Biomarkers, on the contrary, are able to utilize end-points which might give a 'generalised' response. However, in the lack of relevant literature on what response aerosol samples would elucidate on those systems, these assays remain only a possibility.

Acknowledgment

The authors are grateful for the financial support of the grant of the Hungarian Research Fund OTKA K 101484.

References

AFNOR (1991). *Determination de l'inhibition de la luminescence de Photobacterium phosphoreum,*. NFT90-320, Paris.

Andre, M. (2004). The ARTEMIS European driving cycles for measuring car pollutant emissions. *Science of the Total Environment*, 334–335, 73–84.

Arsene, F., Tomoyasu, T. & Bukau, B. (2000). The heat shock response of *Escherichia coli. International Journal of Food Microbiology*, 55, 3–9.

Ayrault, S., Clochiatti, R., Carrot, F., Daudin, L. & Bennett, J.P. (2007). Factors to consider for trace element deposition biomonitoring surveys with lichen transplants. *Science of the Total Environment*, 372, 717–727.

Baptista, M.S., Teresa, M., Vasconcelos, S.D., Cabral, J.P., Freitas, M.C. & Pacheco, A.M.G. Copper, nickel and lead in lichen and tree bark transplants over different periods of time. *Environmental Pollution*, 151, 408-413.

Bergamaschi, L., Rizzio, E., Giaveri, G., Loppi, S. & Gallorini, M. (2007). Comparison between the accumulation capacity of four lichen species transplanted to a urban site. *Environmental Pollution*, 148, 468-476.

Bermudez, G.M.A., Rodriguez, J.H. & Pignata, M.L. (2009). Comparison of the air pollution biomonitoring ability of three Tillandsia species and the lichen *Ramalina celastri* in Argentina. *Environmental Research*, 109, 6–14.

Branquinho, C., Gaio-Oliveira, G., Augusto, S., Pinho, P., Máguas, C. & Correia, C. (2008). Biomonitoring spatial and temporal impact of atmospheric dust from a cement industry. *Environmental Pollution*, 151, 292-299.

Brinkmann, M., Hudjetz, S., Kammann, U., Hennig, M., Kuckelkorn, J., Chinoraks, M., Cofalla, C., Wiseman, S., Giesy, J.P., Schäffer, A., Hecker, M., Wölz, J., Schüttrump, H. & Hollert, H. (2013). How flood events affect rainbow trout: Evidence of a biomarker cascade in rainbow trout after exposure to PAH contaminated sediment suspensions? *Aquatic Toxicology*, 128– 129, 13– 24

Brouwer, H., Murphy, T. & McArdle, L. (1990). A sediment-contact bioassay with *Photobacterium phosphoreum*. *Environmental Toxicology and Chemistry*, 9, 1353-1358.

Brunialti, G. & Frati, L. (2007). Biomonitoring of nine elements by the lichen *Xanthoria parietina* in Adriatic Italy: A retrospective study over a 7-year timespan. *Science of the Total Environment*, 387, 289–300.

Bulich, A. A. (1979). Use of luminescent bacteria for determining toxicity in aquatic environment. In: Markings, L. L., Kimerie, R. A. (Eds.) *Aquatic Toxicology,* ASTM, Philadelphia, pp. 97-105.

Bulich A.A. & Isenberg D.L. (1981). Use of the luminescent bacterial system for the rapid assessment of aquatic toxicity. *ISA Transactions*, 20, 29-33.

Burton, S.A., Peterson, R.V., Dickman, S.N. & Nelson, J.R. (1986). Comparison of in vitro bacterial bioluminescence and tissue culture biomaterials. *Journal of Biomedical Material Research*, 20, 827- 838.

Campisi,T., Abbondanzi, F., Casado-Martinez, C., DelValls, TA., Guerra, R. & Iacondini, A. (2005). Effects of sediment turbidity and color on light output measurements for Microtox Basic Solid-PhaseTest. *Chemosphere*, 60, 9–15.

Carreras, H.A. & Pignata, M.L. (2002). Biomonitoring of heavy metals and air quality in Cordoba City, Argentina, using transplanted lichens. *Environmental Pollution*, 117, 77–87.

Chang, S.T., Lee, H.J. & Gu, M.B. (2004). Enhancement in the sensitivity of an immobilized cell-based soil biosensor for monitoring PAH toxicity. *Sensors and Actuators B,* 97, 272–276.

Chen, K.S., Lin, C.F. & Chou, Y.M. (2001). Determination of source contributions to ambient PM 2.5 in Kaohsiung, Taiwan, using a recept or model. *Journal of the Air and Waste Management Association*, 51, 489-498.

Councell, T.B., Duckenfield, K.U., Landa, E.R. & Callender, E. (2004). Tire-wear particles as a source of zinc to the environment. *Environmental Science and Technology*,38, 4206–4214.

Conti, M.E. & Cecchetti, G. (2001). Biological monitoring: lichens as bioindicators of air pollution assessment. *Environmental Pollution*, 114, 471–92.

Davidov, Y., Rozen, R., Smulski, D.R., VanDyk, T.K., Vollmer, A.C., Elsemore, D.A., LaRossa, R.A. & Belkin, S. (2000). Improved bacterial SOS promoter::lux fusions for genotoxicity detection. *Mutation Research,* 466, 97–107.

Day, K.E., Holtze, K.E., Metcalfe-Smith, J.L., Bishop, C.T. & Dutka, B.J. (1993). Toxicity of leachate from automobile tires to aquatic biota. *Chemosphere*, 27, 665–675.

DeVizcaya-Ruiz, A., Gutiérrez-Castillo, M.E., Uribe-Ramirez, M., Cebrián, M.E., Mugica-Alvarez, V., Sepulvéda, J., Rosas, I., Salinas, E., Garcia-Cuéllar, C., Martínez, F., Alfaro-Moreno, E., Torres-Flores, V., Osornio-Vargas, A., Sioutas, C., Fine, P.M., Singh, M., Geller, M.D., Kuhn, T., Miguel, A.H., Eiguren-Fernandez, A., Schiestl, R.H., Reliene, R. & Froines, J. (2006). Characterization and in vitro biological effects of concentrated particulate matter from Mexico City. *Atmospheric Environment* 40, 583–592.

D'Souza, S.F. (2001). Microbial biosensors. *Biosensors & Bioelectronics*, 16, 337–353.

El-Alawi, Y.S., Dixon, D.G. & Greenberg, B.M. (2001). Effects of a pre-incubation period on the photoinduced toxicity of polycyclic aromatic hydrocarbons to the luminescent bacterium *Vibrio fischeri*. *Environmental Toxicology*, 16, 277-286.

El-Alawi, Y.S., McConkey, B.J., Dixon, D.G. & Greenberg, B.M. (2002). Measurement of short-and long-term toxicity of polycyclic aromatic hydrocarbons using luminescent bacteria. *Ecotoxicology and Environmental Safety*, 51, 12-21.

Elsemore, D.A. (1998). Insertion of promoter region::luxCDABE fusions into the *Escherichia coli* chromosome. *Methods in Molecular Biology*, 102, 97–104.

Eltzov, E., Pavluchkov, V., Burstin, M. & Marks, R.S. (2011). Creation of a fiber optic based biosensor for air toxicity monitoring. *Sensors and Actuators B Chemical*, 155(2), 859-867.

Eom, I.C., Rast, C., Veber, A.M. & Vasseur, P. (2007). Ecotoxicity of a polycyclic aromatic hydrocarbon (PAH)-contaminated soil. *Ecotoxicology and Environmental Safety*, 67, 190–205.

EC, European Commission (2008). Directive 2008/50/EC of the European Parliament and of the Council of 21 May 2008 on ambient air quality and cleaner air for Europe. Technical Report 2008/50/EC, L152 *Off J Eur Comm.*

Fort, F. (1992). Correlation of Microtox EC, with Mouse LD, *In Vitro Toxicology*, 5, 73-82.

Francioni, E., Wagener, A.L.R., Scofield, A.L., Depledge, M.H. & Cavalier, B. (2007). Evaluation of the mussel *Perna perna* as a biomonitor of polycyclic aromatic hydrocarbon (PAH) exposure and effects. *Marine Pollution Bulletin,* 54, 329–338.

Franzle, O. (2006). Complex bioindication and environmental stress assessment. *Ecological Indicators,* 6, 114-136.

Fulladosa, E., Murat, J.C., Martínez, M. & Villaescusa, I. (2005). Patterns of metals and arsenic poisoning in *Vibrio fischeri* bacteria. *Chemosphere*, 60, 43–48.

Gelencsér, A., May, B., Simpson, D., Sánchez-Ochoa, A., Kasper-Giebl, A., Puxbaum, H., Caseiro, A., Pio, C. & Legrand, M. (2007). Source apportionment of PM2.5 organic aerosol over Europe: Primary/secondary, natural/anthropogenic, and fossil/biogenic origin. *Journal of Geophysical Research* 112, D23S04, doi: 10.1029/2006JD008094

Gelencsér, A., Kováts, N., Turóczi, B., Rostási, Á., Hoffer, A., Imre, K., Nyírő-Kósa, I., Csákberényi-Malasics, D., Tóth, Á., Czitrovszky, A., Nagy, A., Nagy, Sz., Ács, A., Kovács, A., Ferincz, Á., Hartyáni. Zs. & Pósfai., M. (2011). The red mud accident in Ajka (Hungary): characterization and potential health effects of fugitive dust. *Environmental Science and Technology, 45* (4), 1608-1615.

George, S.G. (1994). Enzymology and molecular biology of phaseII xenobiotic-conjugating enzymes in fish. In: Mallins, D.C., Ostrander, G.K.(Eds.), *Aquatic Toxicology: Molecular, Biochemical and Cellular Perspectives.* Lewis Publishers, USA, pp.37–85.

Gil, G.-C., Mitchell, R.J., Chang, S.-T. & Gu, M.-B. (2000). A biosensor for the detection of gas toxicity using a recombinant bioluminescent bacterium. *Biosensors and Bioelectronics*., 15, 23–30.

Gil, G.C., Kim, Y.J. & Gu, M.B. (2002). Enhancement in the sensitivity of a gas biosensor by using an advanced immobilization of a recombinant bioluminescent bacterium. *Biosensors and Bioelectronics*,17, 427–432.

Giltrap, M., Macken, A., McHugh, B., Hernan, R., O'Rourke, K., McGovern, E., Foley, B. & Davoren, M. (2009). Bioassay-directed fractionation of marine sediment solvent extracts from the east coast of Ireland. *Chemosphere*, 76, 357–364.

Godinho, R.M., Verburg, T.G., Freitas, M.C. & Wolterbeek, H.Th. (2009). Accumulation of trace elements in the peripheral and central parts of two species of epiphytic lichens transplanted to a polluted site in Portugal. *Environmental Pollution*, 157, 102–109.

Graham, L. 2005. Chemical characterization of emissions from advanced technology light-duty vehicles. *Atmospheric Environment* 39, 2385-2398.

Gu, M.B. & Chang, S.T. (2001). Soil biosensor for the detection of PAH toxicity using an immobilized recombinant bacterium and a biosurfactant. *Biosensors & Bioelectronics*, 16, 667–674.

Gualtieri, M., Andrioletti, M., Mantecca, M., Vismara, C. & Camatini, C. (2005a). Impact of tire debris on in vitro and in vivo systems. *Particle and Fibre Toxicology*, 2:1.

Gualtieri, M., Andrioletti, M., Vismara, C., Milani, M. & Camatini, M. (2005b). Toxicity of tire debris leachates. *Environment International* 31, 723–730.

Gualtieri, M., Mantecca, P., Corvaja, V., Longhin, E., Perrone, M.G., Bolzacchini, E. & Camatini, M. (2009). Winter fine particulate matter from Milan induces morphological and functional alterations in human pulmonary epithelial cells (A549). *Toxicology Letters*, 188, 52-62.

Harkey, G.A. & Young, T.M. (2000). Effect of soil contaminant extraction method in determining toxicity using the Microtox® assay. *Environmental Toxicology and Chemistry,* 19, 276–282.

Hartwell, S.I., Jordahl, D.M. & Dawson, E.O. (2000). The effect of salinity on tire leachate toxicity. *Water, Air and Soil Pollution*, 121, 119–131.

Hirmann, D., Loibner, A.P., Braun, R. & Szolar, O.H.J. (2007). Applicability of the bioluminescence inhibition test in the 96-well microplate format for PAH-solutions and elutriates of PAH-contaminated soils. *Chemosphere*, 67, 1236–1242.

ISO (1998). *Water quality — Determination of the inhibitory effect of water samples on the light emission of Vibrio fischeri* (Luminescent bacteria test) (first edition)

ISO 11348-3:2007, *Water quality — Determination of the inhibitory effect of water samples on the light emission of Vibrio fischeri* (Luminescent bacteria test)

ISO 21338:2010, *Water quality — Kinetic determination of the inhibitory effects of sediment, other solids and coloured samples on the light emission of Vibrio fischeri* (kinetic luminescent bacteria test)

Jennings, V.L.K., Rayner-Brandes, M.H. & Bird, D.J. (2001). Assessing chemical toxicity with the bioluminescent photobacterium (*Vibrio fischeri*): a comparison of three commercial systems. *Water Research*,. 35, 3448-3456

Kim, B.C., Park, K.S., Kim, S.D. & Gu, M.B. (2003). Evaluation of a high throughput toxicity biosensor and comparison with a *Daphnia magna* bioassay. *Biosensors and Bioelectronics*,18, 821-826.

Klumpp, A., Ansel, W., Klumpp, G., Breuer, J., Vergne, P., Sanz , M.J., Rasmussen, S., Ro-Poulsen, H., Artola, A.R., Penuelas, J., He, S., Garrec,J.P., Calatayud, V. (2009). Airborne trace element pollution in 11 European cities assessed by exposure of standardised ryegrass cultures. *Atmospheric Environment*, 43, 329–339.

Komori, K., Miyajima, S., Tsuru, T., Fujii, T., Mohri, S., Ono, Y. & Sakai, Y. (2009). A rapid and simple evaluation system for gas toxicity using luminous bacteria entrapped by a polyion complex membrane. *Chemosphere*, 77, 1106–1112.

Kovács, M. (ed). (1992). *Biological indicators in environmental protection.* Academic Publisher, Budapest

Kováts, N., Kovács, A., Ács, A., Ferincz, Á., Turóczy, B.& Gelencsér, A., (2012). Development of a whole-aerosol test protocol. *Environmental Toxicology and Pharmacology*, 33, 284–287.

Kováts, N., Ács, A., Ferincz, Á., Kovács, A., Horváth, E., Kakasi, B., Jancsek-Turóczi, B., Gelencsér, A. (2013). Ecotoxicity and genotoxicity assessment of exhaust particulates from diesel-powered buses. *Environmental Monitoring and Assessment*, doi:10.1007/s10661-013-3206-3

Lappalainen, J., Juvonen, R., Vaajasaari, K. & Karp, M. (1999). A new flash method for measuring the toxicity of solid and colored samples. *Chemosphere*, 38, 1069-1083.

Lappalainen, J., Juvonen, R., Nurmi, J. & Karp, M. (2001). Automated color correction method for *Vibrio fischeri* toxicity test. Comparison of standard and kinetic assays. *Chemosphere*, 45, 635-641.

Lebsack, M. E., Anderson, A.D. DeGraeve, G.M. & Bergman, H.L. (1981). Comparison of Bacterial Luminescence and Fish Bioassay Results for Fossil-Fuel Process Waters and Phenolic Constituents. In: Branson, D.R., and Kickson, K. L. (Eds). Aquatic Toxicology and Hazard Assessment: Fourth Conference, ASTM STP 737, *American Society for Testing and Materials,* 348-356.

Lee, H.J., Villaume, J., Cullen, D.C., Kima, B.C. & Gu, M.B. Monitoring and classification of PAH toxicity using an immobilized bioluminescent bacteria. *Biosensors and Bioelectronics*,18, 571-577

Lin, T.C. & Chao, M.R. (2002). Assessing the influence of methanol-containing additive on biological characteristics of diesel exhaust emissions using Microtox and Mutatox assays. *The Science of the Total Environment*, 284, 61-74.

Mantecca, P., Gualtieri, M., Andrioletti, M., Bacchetta, R., Vismara, C., Vailati, G. & Camatini, M. (2007). Tire debris organic extract affects *Xenopus development. Environment International*, 33, 642–648.

Marigómez, I., Soto, M., Cancio, I., Orbea, A., Garmendia, L. & Cajaraville, M.P. (2006). Cell and tissue biomarkers in mussel and histopathology in hake and anchovy from Bay of Biscay after the Prestige oil spill (Monitoring Campaign 2003). *Marine Pollution. Bulletin* 53, 287–304.

Markert, B. (2007). Definitions and principles for bioindication and biomonitoring of trace metals in the environment. *Journal of Trace Elements in Medicine and Biology*, 21 S1, 77-82.

Marques, A.P., Freitas, M.C., Reis, M.A., Wolterbeek, H.Th., Verburg, T. & de Goeij, J.J.M. (2004). Lichen-transplant biomonitoring in the assessment of dispersion of atmospheric trace-element pollutants: effects of orientation towards the wind direction. *Journal of Atmospheric Chemistry*, 49, 211–222.

Marwood, C., McAtee, B., Kreider, M., Ogle, R.S., Finley, B., Sweet, L. & Panko, J. (2011). Acute aquatic toxicity of tire and road wear particles to alga, daphnid, and fish. *Ecotoxicology*, 20, 2079–2089.

Microbics Corporation (1992). Microtox manual. A Toxicity Testing Handbook.

Moriarty, F. (1983). *Ecotoxicology*. Second Edition. Academic Press, Harcourt Brace Jovanovich Publishers.

Mortimer, M., Kasemets, K., Heinlaan, M., Kurvet, I. & Kahru, A. (2008). High throughput kinetic *Vibrio fischeri* bioluminescence inhibition assay for study of toxic effects of nanoparticles. *Toxicology in Vitro*, 22, 1402-1417.

Nahrgang, J. Camus, L., Gonzalez, P., Goksøyr, A., Christiansen, J.B. & Hope, H. (2009). PAH biomarker responses in polar cod (*Boreogadus saida*) exposed to benzo(a)pyrene. *Aquatic Toxicology,* 94, 309–319.

Nelson, S.M., Mueller, G. & Hemphill, D.C. (1994). Identification of tire leachate toxicants and a risk assessment of water quality effects using tire reefs and canals. *Bulletin of Environmental Contamination and Toxicology*, 52, 574–581.

Ocampo-Duque, W., Sierra, J., Ferré-Huguet, N., Schuhmacher, M. & Domingo, J.L. (2008). Estimating the environmental impact of micropollutants in the low Ebro River (Spain): an approach based on screening toxicity with *Vibrio fischeri*. *Chemosphere*, 72, 715–21.

Olajire, A., Altenburger, R., Küster, E. & Brack, W. (2005). Chemical and ecotoxicological assessment of polycyclic aromatic hydrocarbon—

contaminated sediments of the Niger Delta, Southern Nigeria. *Science of the Total Environment,* 340,123–136.

Pape-Lindstrom, P. A. & Lydy, M.J. (1997). Synergistic toxicity of atrazine and organophosphate insecticides contravenes the response addition mixture model. *Environmental Toxicology and Chemistry*. 1997, 16, 2415–20.

Peakall, D. (1992). *Animal Biomarkers as Pollution Indicators*. Chapman & Hall, London.

Preuss, R., Angerer, J.& Drexler, H. (2003). Naphthalene—an environmental and occupational toxicant. *International Archives for Occupational and Environmental Health*, 76, 556–576.

Ringwood, A.H, DeLorenzo, M.E, Ross, P.E & Holland, A.F. (1997). Interpretation of the Microtox solid-phase toxicity tests: the effect of sediment composition. *Environmental Toxicology and Chemistry*, 16, 1135–40.

Ripp, S., Nivens, D.E., Werner, C. & Sayler, G.S. (2000). Bioluminescent most-probable-number monitoring of a genetically engineered bacterium during a long-term contained field release. *Applied Microbiology and Biotechnology,* 53, 736–741.

Roig, N., Nadal, M., Sierra, J., Ginebreda, A., Schuhmacher, M. & Domingo, J L. (2011). Novel approach for assessing heavy metal pollution and ecotoxicological status of rivers by means of passive sampling methods. *Environment International*, 37, 671–677.

Roig, N., Sierra, J., Rovira, J., Schuhmacher, M., Domingo, J.L. & Nadal, M. (2013). In vitro tests to assess toxic effects of airborne PM10 samples. Correlation with metals and chlorinated dioxins and furans. *Science of the Total Environment*, 443, 791–797.

Rusu, A.-M., Jones, G.C., Chimonides, P.D.J. & Purvis, O.W. (2006). Biomonitoring using the lichen *Hypogymnia physodes* and bark samples near Zlatna, Romania immediately following closure of a copper ore-processing plant. *Environmental Pollution*, 143, 81-88

Salizzato, M., Pavoni, A.P., Ghirardini, A.V. & Ghetti, P.F. (1998). Sediment toxicity measured using Vibrio fischeri as related to the concentrations of organic (PCBs, PAHs) and inorganic (metals, sulphur) pollutants. *Chemosphere*, 36, 2949-2968.

Sayler, G.S., Cox, C.D., Burlage, R., Ripp, S., Nivens, D.E., Werner, C., Ahn, Y. & Matrubutham, U. (1999). Field application of a genetically engineered microorganism for polycyclic aromatic hydrocarbon bioremediation process monitoring and control. In: Fass, R., Flashner, Y.,

Reuveny, S. (Eds.), *Novel Approaches for Bioremediation of Organic Pollution*. Kluwer Academic/Plenum Press, NewYork, pp.241–254.

Schauer, J. J., Rogge, W. F., Hildemann, L. M., Mazurek, M. A. & Cass, G. R. (1996). Source apportionment of airborne particulate matter using organic compounds as tracers. *Atmospheric Environment*, 30 (22), 3837–3855.

Schlenk, D., Celander, M., Gallagher, E.P., George, S., James, M., Kullman, S.W., van der Hurk, P. & Willett, K. (2008). Biotransformation in fishes. In: DiGiulio, R.T., Hinton, D.E. (Eds.), *The Toxicology of Fishes*. CRC Press, Taylor & Francis Group, pp.153–234.

Schultz, T.W., Cronin, M.T.D., Walker, J.D. & Aptula, A.O. (2003).Quantitative structure-activity relationship (QSARs) in toxicology: a historical perspective. *Journal of Molecular Structure-Theochem*, 622, 1–22

Szczepaniak, K. & Biziuk, M. (2003). Aspects of the biomonitoring studies using mosses and lichens as indicators of metal pollution. *Environmental Research*, 93, 221–230.

Suter, G.W. (1993). *Ecological risk assessment.* Lewis Publishers.

Takada, H., Onda, T., Harada, M., Ogura, N. (1991). Distribution and sources of polycyclic aromatic hydrocarbons (PAHs) in street dust from the Tokyo Metropolitan area. *Science of the Total Environment*, 107, 45–69.

Tretiach, M., Adamo, P., Bargagli, R., Baruffo, L., Carletti, L., Crisafulli, P., Giordano, S., Modenesi, P., Orlando, S. & Pittao, E. (2007). Lichen and moss bags as monitoring devices in urban areas. Part I: Influence of exposure on sample vitality. *Environmental Pollution*,146, 380-391.

Triolo, L., Binazzi, A., Cagnetti, P., Carconi, P., Correnti, A., DeLuca, E., DiBonito, R., Grandoni, G., Mastrantonio, M., Rosa, S., Schimberni, M., Uccelli, R. & Zappa, G. (2008). Air pollution impact assessment on agroecosystem and human health characterisation in the area surrounding the industrial settlement of Milazzo (Italy): a multidisciplinary approach. *Environmental Monitoring and Assessment*, 140, 191–209.

Tung, K.K., Scheibner, B., Miller, T. & Bulich, A.A. (1990*). A new method for testing soil and sediment samples.* Presented at the SETAC Conference, November 1990.

Turóczi, B., Hoffer, A., Kováts, N., Ács, A., Tóth, Á. & Gelencsér, A. (2012). Comparative assessment of ecotoxicity of urban aerosol. *Athmospheric Chemistry and Physics*, 12, 7365–7370.

USEPA (2000*). Method Guidance and Recommendations for Whole Effluent Toxicity (WET) Testing*. 40 CFR Part 136, EPA 821-B-00-004

Valdman, E., Valdman, B., Battaglini, F. & Leite, S.G.F. (2004). On-line detection of low naphthalene concentrations by a bioluminescent sensor. *Process Biochemistry*, 39, 1217–1222.

Valdman, E., Gutz, I.G.R. (2008). Bioluminescent sensor for naphthalene in air: Cell immobilization and evaluation with a dynamic standard atmosphere generator. *Sensors and Actuators*, B133, 656–663.

Varga, R, Kibedi-Varga, L, Markovits-Somogyi, R, Torok, A, Meszaros, F. (2010). Statistical assessment of traffic quality in Budapest. In: Borkowski, S., Nabialek, M. (Eds.) *Toyotarity: Knowledge using in production management.* Dnepropetrovsk: Yurii V Makovetsky, 2010. pp. 143-154.

Vassault, A. (1983). Lactate dehydrogenase. In: Bergmeyer, M.O.(Ed.), *Methods of Enzymatic Analysis, Enzymes: Oxireductases, Transferases,* vol.3. Academic Press, New York, pp.118–126.

Vollmer, A.C., Belkin, S., Smulski, D.R., VanDyk, T.K. & LaRossa, R.A. (1997). Detection of DNA damage by use of *Escherichia coli* carrying recA::lux, uvrA::lux, or alkA::lux reporter plasmids. *Applied and Environmental Microbiology,* 63, 2566–2571.

Vouitsis, E., Ntziachristos, L., Pistikopoulos, P., Samaras, Z., Chrysikou, L., Samara, C., Papadimitriou, C., Samaras, P. & Sakellaropoulos, G. (2009). An investigation on the physical, chemical and ecotoxicological characteristics of particulate matter emitted from light-duty vehicles. *Environmental Pollution,* 157, 2320–2327.

Walker, C.H., Hopkin, S.P., Sibly, R.M. & Peakall, D.B. (2006). *Principles of ecotoxicology.* Third edition. Taylor and Francis, Boca Raton

Wik, A. & Dave, G. (2005). Environmental labeling of car tires toxicity to *Daphnia magna* can be used as a screening method. *Chemosphere*, 58, 645–651.

Wik, A. & Dave, G. (2006). Acute toxicity of leachates of tire wear material to *Daphnia magna* - Variability and toxic components. *Chemosphere*, 64, 1777–1784.

Wik, A. & Dave, G. (2009). Occurrence and effects of tire wear particles in the environment – A critical review and an initial risk assessment. *Environmental Pollution*, 157, 1–11.

Wolterbeek, B. (2002). Biomonitoring of trace element air pollution: principles, possibilities and perspectives. *Environmental Pollution*, 120, 11-21.

WHO, World Health Organization (2006). Health risks of particulate matter from long-range transboundary air pollution. *Joint WHO/Convention Task Force on the Health Aspects of Air Pollution.* Copenhagen, Denmark.

In: Air Quality
Editor: Arthur Hermans

ISBN: 978-1-62808-259-3

Chapter 5

Air Quality in Occupied School Building Spaces in the South of Portugal

Eusébio Z. E. Conceição[1] and M. Manuela J. R. Lúcio[1]
[1]Faculty of Sciences and Technology, University of Algarve, Campus of Gambelas, Faro, Portugal

Abstract

The aim of this paper is to evaluate the air quality in occupied school buildings spaces in the South of Portugal. The results of this review article are obtained in school buildings, with Mediterranean environment, from Kindergartens to University. In this work the numerical and the experimental analysis are presented and discussed.

In the numerical simulation the softwares, that simulate the building thermal behaviour and the computational fluid dynamics, are used. In the building thermal behaviour software the mean indoor air quality inside all buildings occupied spaces is evaluated, while in the computational fluid dynamics numerical model the air quality inside each occupied space, and mainly in the respiration area, is evaluated in detail. The building thermal behaviour numerical model considers the building opaque and transparent bodies, while the computational fluid dynamics numerical model considers, in each space, the occupants and the indoor bodies.

In the experimental measurements the carbon dioxide is evaluated. The carbon dioxide, using the tracer gas decreasing concentration, is used to evaluate the internal air quality, using the air exchange rate, the age of the air or the airflow rate.

The numerical and experimental techniques are used to evaluate the internal air quality level in spaces occupied by children and students, in the South of Portugal. The obtained results are used to develop techniques and methodologies in order to increase the internal air quality in these kinds of occupied spaces. In the study the natural and crossed ventilation, in cold and warm thermal conditions, are evaluated.

Keywords: School building spaces, numerical values, experimental results, air quality

INTRODUCTION

In the South of Portugal, in the Algarve region, a climate with high solar radiation levels, mainly in Summer conditions, but also in the other seasons, is expected. The external air temperature in the Summer conditions is warm, in the Autumn conditions and in the Spring conditions is light warm and in the Winter conditions is not very cold.

The occupied school buildings spaces in the south of Portugal are subjected not only to the solar radiation level and to the external air temperature levels, but also to the air renovation level. All these environmental variables influence the building thermal response and the indoor thermal environment, however the air renovation level also influences the indoor air quality.

The air renovation level is associated to the used ventilation process. In the South of Portugal, in general, the school buildings air renovation is based in renewable energies methodologies. In general the crossed ventilation, in parallel walls and in adjacent walls, with inlet air temperature equal to the outlet air temperature, applied mainly in Autumn and Spring conditions, is used. However, in Winter conditions the greenhouse, located in the roof and in the hall, and in Summer conditions the underground ducts are applied. New ventilation systems, as personalized ventilation with one and two air terminal devices, that can be used in the near future, are also analysed in detail.

The main objective of this study is to present a review article, of works developed in the South of Portugal, in Winter, Spring, Summer and Autumn conditions, in school buildings with Mediterranean environment, from Kindergartens to University. The idea is to propose some improvements for the future construction in this region, with the aim to increase the indoor air quality with low energy consumption level. In this paper the numerical and the experimental analysis will also be applied.

RESULTS AND DISCUSSION

In this section the crossed ventilation, the greenhouse, the underground ducts and the personalized ventilation systems are analysed. In the crossed ventilation, the crossed ventilation in parallel walls using upper airflow, the crossed ventilation in parallel walls using upwards airflow and the crossed ventilation in adjacent walls using upper airflow are analysed. In the greenhouse, the greenhouse located in the roof and the greenhouse located in the hall are analysed.

Finally, in the personalized ventilation the personalized ventilation with one air terminal device and the personalised ventilation with two air terminal devices are analysed.

Crossed Ventilation in Parallel Walls Using Upper Airflow

In this section the crossed ventilation in parallel walls using upper airflow is analysed. This philosophy considers that the air inlet is done above the head level, in a lateral wall, and the air outlet is done above the head level, in another parallel lateral wall. The contaminants released by the occupants, located between the inlet and the outlet walls, are transported by the airflow to the outlet area.

Conceição et al. (2007a) and Conceição et al. (2008a) presented detailed studies about the indoor thermal environment in a classroom equipped with this kind of crossed ventilation.

Conceição et al. (2007a), using an experimental study in a classroom, evaluated the internal air quality levels that students are subjected. The airflow inlet in the doors' lower and upper area and the outlet located in the windows was studied. The experimental test was made in Summer conditions, in a warm day, using the inlet airflow coming from the corridor compartment. In accordance with the obtained results, when the inlet airflow made in the upper area, the air renovation flow rate did not guarantee acceptable indoor air quality.

In Conceição et al. (2008a), in a moderated Summer typical day, the air quality, was evaluated. In this study three experimental tests were made:

- in the first test the curtains were open, the four small windows were open and the main windows and door were closed;
- in the second test the curtains were closed, the four small windows were open and the main windows and door were closed;
- in the third test the curtains were open, the four small windows were closed and the main windows and door were open.

In accordance with the obtained results, in the first analysed test the air renovation in the classroom central breathing area was higher than the mean air renovation level. The first test guaranteed the best air quality levels in the breathing area. In the second test, the air renovation rate in the classroom central breathing area was lightly lower than the mean air renovation level. The air quality levels, obtained in the tests, were very close to the ANSI/ASHRAE Standard 62.1 (2004) suggestions. Nevertheless, in accord to the obtained results, the introduction of indoor curtains placed above the small windows levels was suggested.

Crossed Ventilation in Parallel Walls Using Upwards Airflow

In this section the crossed ventilation in parallel walls using upwards airflow is analysed. This philosophy considers that the air inlet is done in the feet level, in a lateral wall, and the air outlet is done above the head level, in another parallel lateral wall. The contaminants released by the occupants, located between the inlet and the outlet walls, are transported by the airflow to the outlet area.

Conceição and Lúcio (2006), Conceição et al. (2007a) and Conceição and Lúcio (2009) presented detailed studies about the air quality inside compartments of a school building, using the air exchange and the carbon dioxide concentration for different ventilation strategies.

In Conceição and Lúcio (2006), using the tracer gases method in a classroom, the air exchange and the flow rate were experimentally obtained. In accord to the obtained results was verified that the crossed ventilation system implemented in the building's compartment increase the airflow rate. The evolution of carbon dioxide, inside the different spaces with different air flow typologies, was numerically calculated. It was verified that the carbon dioxide concentration is higher than the recommended value presented in ANSI/ASHRAE Standard 62.1 (2004). A new methodology, using a forced ventilation philosophy with and without air flow rate adjustment, was implemented. Without air flow adjustment was verified that the classrooms are subjected to acceptable carbon dioxide concentration and the corridors, atria and small offices, are subjected to non-acceptable carbon dioxide concentration. However, with air flow adjustment, acceptable carbon dioxide concentration in all spaces was verified.

In Conceição et al. (2007a), in an experimental study made inside a classroom, when the inlet airflow made in the doors' lower and upper area and the outlet is made in the wall, the local airflow rate was lightly lower when the ventilator is placed in the door's lower area than when the ventilator is placed in the door's upper area. However, in both situations the air renovation flow rate did not guarantee acceptable indoor air quality. In Conceição et al. (2007a) is suggested to use more than one ventilator, as extractor way, in the windows upper area.

In Conceição and Lúcio (2009) a numerical model that simulates the school thermal behaviour and evaluates the indoor thermal and air quality, in transient conditions, for four different orientations, either in winter or summer conditions, were made. A new airflow topology, in order to guarantee acceptable indoor air quality, was developed. This airflow topology considered the inlet airflow in the upper area door corridors, the airflow passage from the corridor to the classrooms and the outlet airflow in the upper area classroom windows.

Crossed Ventilation in Adjacent Walls Using Upper Airflow

In this section the crossed ventilation in adjacent walls using upper airflow is analysed. This philosophy considers that the air inlet is done above the head level and the air outlet is done above the head level, in an adjacent wall. The contaminants released by the occupants, located inside the space, are transported by the airflow to the outlet area.

Conceição et al. (2008b) presented detailed studies about the airflow inside school buildings office compartments with moderate environment. A computational fluid dynamics and a full-scale experimental chamber were developed and used in the study. The carbon dioxide concentration evolution in the respiration area and other environmental mean values in the occupied area were measured. In accordance with the obtained results was verified that the airflow topology is divided into two zones: a main airflow in the upper non-occupied area and an airflow in the lower occupied area. In the non-occupied area the inlet and the outlet airflow were placed. In accordance with the obtained results, the measured airflow rate means value guaranteed acceptable indoor air quality.

Greenhouse Located in the Roof

In this section the greenhouse located in the roof area is analysed. This philosophy considers that the air inlet is done in the greenhouse and the air outlet is done in the occupied spaces. The airflow, transported by a ducts system, is responsible to guarantee acceptable thermal conditions and acceptable indoor air quality in the classrooms.

Conceição and Lúcio (2008a) and Conceição and Lúcio (2008b) presented detailed studies about the thermal study of school buildings, equipped with greenhouse located in the roof, in winter conditions. In this study the indoor thermal quality and the indoor air quality were simultaneously analysed. The occupied spaces were ventilated from the collector, from the corridor and from the external environment. The indoor air quality in transient conditions was evaluated.

In accordance with the obtained results, it was verified that the solar air collectors, placed in the roof, guarantee acceptable air quality levels in the occupied spaces.

Greenhouse Located in the Hall

In this section the greenhouse located in the hall is analysed. This philosophy considers that the air inlet is done in the greenhouse and the air outlet is done in the occupied spaces. The airflow, transported by a ducts system, is responsible to guarantee acceptable thermal conditions and acceptable indoor air quality.

Conceição et al. (2008c) and Conceição and Lúcio (2010) presented detailed studies about the application of an indoor greenhouse, namely the implementation of passive and active solar strategies in a kindergarten in Mediterranean external conditions.

In Conceição et al. (2008c), in a preliminary study, a numerical study, that simulates the kindergarten thermal response and evaluates the indoor thermal comfort and air quality, in transient conditions, was used. In the study was suggested to consider the hall, as greenhouse, used in the ventilation of the classrooms with window turned North. It is also suggested to use a duct system, or the corridor, to transport the airflow from the greenhouse to the cold classrooms with windows turned North.

In Conceição and Lúcio (2010) the continuation of the previous numerical work was presented. A real occupation cycle, the compartments, the building opaque bodies, the building transparent bodies and the external shading devices were considered. The natural and forced ventilation, in Summer and in Winter conditions, were applied and the passive and active strategies were developed. In Summer conditions, the cold forced airflow during the night coming from the external environment and during the day coming from the underground space. In Winter conditions the warm forced airflow coming from an internal greenhouse (see Conceição et al., 2008c). The combination with the passive and active developed strategies guaranteed, in general, acceptable thermal comfort and air quality conditions in the occupation spaces with low energy consumption level.

Underground Ducts

In this section the underground ducts located below the building area are analysed. This philosophy considers that the air inlet is done in the underground ducts and the air outlet is done in the occupied spaces. The airflow, transported by a ducts system, is responsible to guarantee acceptable thermal conditions and acceptable indoor air quality.

In Conceição et al. (2008d) a forced ventilation system in kindergarten spaces was analysed. The airflow coming from the underground ducts, inlet in the classroom floor level, with low air velocity, and outlet above the head level. In the study the computational fluids dynamics was used. In accordance with the obtained results, good indoor air quality was guaranteed for the children and teachers breathing level.

Personalized Ventilation with One Air Terminal Device

In this section the personalized ventilation with one air terminal device is analysed. This philosophy considers that the air inlet is done in front to the occupant, above the desk writing level, and the air outlet is done above the head level, in the ceiling level. The contaminants released by the occupants, are transported by the airflow to the outlet area.

Conceição et al. (2004) and Conceição et al. (2007b) presented detailed studies about the experimental and numerical study of personalized air ventilation in classrooms' desks.

In Conceição et al. (2004), an analysis of an occupant thermal response, subjected to direct solar radiation and subjected to cold airflow, from an air-conditioning system, was analysed. The occupant was simulated by a thermal-manikin, equipped with heating, circulation, control, transpiration and respiration systems. Measurements of air velocity fluctuations around the occupant were made. The manikin's respiration system was analysed in the simulated situation and it was possible to conclude that the exhalation through the nose is disrupted by the air-conditioning system airflow, while exhalation through the mouth is not. These results did not have significant influence in the thermal response of the occupant, but can be very important in terms of air quality exposure.

In Conceição et al. (2007b) an experimental and numerical study of personalized ventilation in classrooms' desks was made. One thermal manikin, one ventilated desk, a multi-nodal human thermal comfort model and two interior climate analysers, were used. In this study with one air terminal devices, placed above the writing area of the desk, was verified a good distribution of air velocity and temperature field around the manikin.

Personalised Ventilation with Two Air Terminal Devices

In this section the personalized ventilation with two air terminal devices is analysed. This philosophy considers that the air inlet is done in front to the occupant, above and below the writing desk level, and the air outlet is done above the head level, in the ceiling level. The contaminants released by the occupants, are transported by the airflow to the outlet area.

Conceição et al. (2007b), Conceição et al. (2008e) and Conceição et al. (2010) presented detailed studies about the experimental and a numerical study of personalized ventilation in classrooms' desks equipped with upper and lower air terminal devices.

In Conceição et al. (2007b), experimental and numerical study of personalized air ventilation in classrooms' desks were analysed. One thermal manikin, one ventilated desk and two interior climate analysers, were used in the experimental setup, while a human thermal comfort model was used in the numerical simulation. In this study, with two air terminal devices, an air velocity uniform field around the manikin, better than the configuration with only one air terminal device, was guaranteed.

In Conceição et al. (2008e), the design used in the configuration of the air terminal devices, installed in the classroom desk, a relatively uniform air velocity field around the manikin was guaranteed. The study was made with and without small ventilators placed in the exit air terminal devices. In accordance with the obtained results, was suggested to introduce grids in the air terminal devices exit area and to lightly change the exit air direction in the air terminal device located above the writing desk area.

In Conceição et al. (2010a) the air quality levels in a classroom space with desks equipped with two personalized ventilation system, was evaluated. An experimental wood chamber equipped with a seated manikin and with a classroom desk and was used. The combination of forced convection, from the air terminal devices, and of free convection, from the thermal manikin, promoted an ascendant airflow around the occupant with highest air renovation rate in the respiration area and with acceptable air quality level in the breathing area.

REFERENCES

ANSI/ASHRAE Standard 62.1 (2004) Ventilation for acceptable indoor air quality, American Society of Heating, Refrigerating and Air-Conditioning Engineers, Inc, 2004.

Conceição E. Z. E., Lourenço T. M. C., Brito A. I. P. V. and Lúcio M. M. J. R. (2004). Evaluation of Human Thermal Response in Occupied Spaces Subjected to Direct Solar Radiation, RoomVent′2004 - 9th International Conference on Air Distribution in Rooms, Coimbra, 5 to 8 de September de 2004.

Conceição E. Z. E.,. Lúcio M. M. J. R, Rosa S. P., Custódio A. L. V., Andrade R. L. and Meira M. J. P. A. (2010). Evaluation of comfort level in desks equipped with two personalized ventilation systems in slightly warm environments", *Building and Environment*, 45 (3), 601-609.

Conceição, E. Z. E. and Lúcio, M. M. J. R. and Lopes M. C. (2008c). Application of an Indoor Greenhouse in the Energy and Thermal Comfort Performance in a Kindergarten School Building in the South of Portugal in Winter Conditions", *WSEAS Transactions on Environment and Development,* 4 (8), 644-654.

Conceição, E. Z. E. and Lúcio, M. M. J. R. and Lopes M. C. (2008d). Ventilação em Salas de Jardins-de-Infância, 1ª Conferências sobre Edifícios Eficientes, Faro, Portugal, 25 of January de 2008.

Conceição, E. Z. E. and Lúcio, M. M. J. R. (2008a). Thermal Study of School Buildings in Winter Conditions", *Building and Environment*, 43 (5), 782-792.

Conceição, E. Z. E. and Lúcio, M. M. J. R. (2008b). Projecto de um Sistema de Ventilação Eficiente de um Edifício Escolar em Condições de Inverno na Região do Algarve, Revista INGENIUM, IIª Série, *Ordem dos Engenheiros,* 107, 76-80.

Conceição, E. Z. E. and Lúcio, M. M. J. R. (2009). Numerical Study of the Thermal Efficiency of a School Building with Complex Topology for Different Orientations, *Indoor and Building Environment*, 18 (1), 41-51.

Conceição, E. Z. E. and Lúcio, M. M. J. R. (2010). Implementation of Passive and Active Solar Strategies in a Kindergarten in Mediterranean External Conditions", *Building Simulation*, 3, 245–261.

Conceição, E. Z. E. and Lúcio, M. M.J. R. (2006). Air Quality Inside Compartments of a School Building: Air Exchange Monitoring, Evaluation of Carbon Dioxide and Assessment of Ventilation Strategies, *The International Journal of Ventilation*, 5 (2), 259-270.

Conceição, E. Z. E., Lúcio, M. M. J. R. and Farinho J. P. (2007b). Experimental and Numerical Study of Personalized Air Ventilation in Classrooms' Desks, RoomVent'2007 - 10th International Conference on Air Distribution in Rooms, Helsinki, Finland, 13 to 15 de June de 2007.

Conceição, E. Z. E., Lúcio, M. M. J. R., Rosa, S. P., Custódio, A. L. V., Meira M. J. P. A. and Andrade, R L. (2008e). Evaluation of comfort level in seated students in classrooms' desks using a personalized ventilation system, Indoor Air 2008, Copenhagen, Denmark, 17 to 22 de August de 2008.

Conceição, E. Z. E., Lúcio, M. M.J. R. and Vicente V. D. S. R. (2007a). Indoor Thermal Environment in a Classroom Equipped with Air Forced System, REHVA World Congress CLIMA'2007 WellBeing Indoors, Helsinki, Finland, 10 to 14 de June de 2007.

Conceição, E. Z. E., Lúcio, M. M.J. R., Vicente V. D. S. R and Rosão V. C. T. (2008a). Evaluation of Local Thermal Discomfort in a Classroom Equipped with Crossed Ventilation, *The International Journal of Ventilation,* 7 (3), 267-277.

Conceição, E. Z. E., Vicente V. D. S. R and Lúcio, M. M.J. R. (2008b). Airflow Inside School Buildings Office Compartments With Moderate

Environment, International Journal on Heating Air Conditioning and Refrigerating Research, ASHRAE, American Society of Heating, *Refrigerating and Air-Conditioning Engineers, Inc.*, 14 (2), 195-207.

INDEX

A

B

C

D

E

F

G

H

I

K

N

O

P

Q

R

S

T

U

V

W

Z